Systems Approach to Astrobiology

Systems thinking/analysis is widely applied for solving complex problems in engineering and certain other fields. Astrobiology, which inherently involves complex problems, can benefit from such an approach. This book provides the background and methodology of this approach for professionals, upper-level undergraduate students, and others with an interest in astrobiology topics.

In addition, this book constitutes a valuable resource not only for astrobiology but also for its foundational disciplines, e.g., chemistry, physics, astronomy, biology, biochemistry, geology, and planetary geology.

Features

- Surveys of the systems approach to analyzing and understanding multifaceted, complex problems in astrobiology, written by two scientists who also have engineering backgrounds.
- Systems applications to areas important to astrobiology, such as chemical evolution, prebiotic chemistry, geochemical/geophysical settings conducive to emergence of life, robotic space exploration, and much more.
- Wide appeal for all readers interested in the origin and occurrence of life in our Solar System and beyond.

Vera M. Kolb earned her BS in Chemical Engineering and MS in Organic Chemistry from Belgrade University, and PhD in Organic Chemistry from Southern Illinois University at Carbondale. She is Professor Emerita of Chemistry at the University of Wisconsin-Parkside (UW-P). During her research at the NASA Specialized Center of Research and Training (NSCORT), she collaborated with Leslie Orgel (Salk Institute) and Stanley Miller (UCSD). In 1992, she received the UW-P Award for Excellence in Research and Creative Activity. She later researched sugar-silicates in a prebiotic context with Joseph Lambert (Northwestern University). In 2002, she was inducted into the Southeastern Wisconsin Educators' Hall of Fame. She has over 150 publications, patents, and books and has received numerous research awards from the Wisconsin Space Grant Consortium/NASA. Prof. Kolb is editor/contributor for several books, including *Astrobiology, An Evolutionary Approach* (CRC, 2015), the 54-chapter *Handbook of Astrobiology* (CRC, 2019), and co-authored with Benton Clark *Astrobiology for a General Reader: A Questions and Answers Approach* (CSP, 2020).

Benton C. Clark III earned his PhD in Biophysics from Columbia University and a Master's in Nuclear Physics and Engineering from the University of California, Berkeley. He is currently Senior Scientist at the Space Science Institute. In 1976, the XRFS instrument he designed for the Viking missions discovered the element composition and high salt content of martian soils. His 40-plus years of experience formerly working with engineers at Lockheed Martin aerospace spanned development of more than a dozen spacecraft, including Mars orbiters and landers, missions to Venus, Jupiter, and the Moon, and sample return missions to comet Wild-2, L1 solar wind, and asteroid Bennu (for which he led design of the sampling system). He is also on the science teams for all four NASA rover missions to Mars (Spirit, Opportunity, Curiosity, and Perseverance). His extensive publications include over two dozen in the journals *Science* and *Nature*.

Systems Approach to Astrobiology

Vera M. Kolb and Benton C. Clark III

CRC Press
Taylor & Francis Group
Boca Raton London New York

CRC Press is an imprint of the
Taylor & Francis Group, an **informa** business

First edition published 2023
by CRC Press
4 Park Square, Milton Park, Abingdon, Oxon, OX14 4RN

and by CRC Press
6000 Broken Sound Parkway NW, Suite 300, Boca Raton, FL 33487-2742

CRC Press is an imprint of Informa UK Limited

British Library Cataloguing-in-Publication Data
A catalogue record for this book is available from the British Library

ISBN: 9781032127149 (hbk)
ISBN: 9781032116280 (pbk)
ISBN: 9781003225874 (ebk)

DOI: 10.1201/9781003225874

Typeset in Times
by codeMantra

VMK dedicates this book to her sister-in-law Mira Kolb and her niece Nataša Kolb, and to the loving memory of her parents, Dr. Martin and Dobrila Kolb, and her brother, Vladimir Kolb.

Contents

Preface

During my teaching career as a Professor of Chemistry, I continued my lifelong interest in thinking methods, and how they can be improved by learners at all levels. For example, I studied misconceptions in students' learning of chemistry topics. I have also attended many educational conferences and talked to many of my colleagues, both teachers and researchers about the ways students think while they learn. Eventually, I adopted a new teaching approach that was aimed toward a better understanding of the subjects in science and other fields, which departed from the traditional reductionist thinking and assumed a more holistic one.

I was, like many other scientists, well trained in the reductionist method. While I understood the need for the holistic approach, I could not grasp the holism from the practical point of solving scientific problems. For example, I was familiar with the Gaia hypothesis, according to which the living forms on the Earth collectively regulate and enable the continuation of life, and thus the biosphere is likened to a vast self-regulating organism. This hypothesis is holistic, and I understood that, and I also believe that this hypothesis is correct. However, I could not grasp how I can practically study the entire Earth's biosphere and the regulation that is going on. Some of my colleagues in the biological field promoted complex thinking which, they told me, is well suited for holistic problems. I tried very hard to understand what complex thinking is, but I could not find a solid, clear, and practical explanation. I gradually learned how to adopt complex thinking mostly via my projects in the field of Green Chemistry, which has 12 principles that must be considered simultaneously. Still, it was challenging to teach complex thinking.

It is difficult to say how long my grappling with the problem of complex thinking would have continued and how many dead ends I would have run into. Luckily for me, I had a great breakthrough, which came from some excellent books on holistic thinking, and then, I hit a jackpot, when I discovered systems thinking. I devoured the material like a starved person, and immediately decided to apply it to astrobiology, in which such an application was either non-existing or sporadic. Quickly, I discovered that many frustrating complex problems of astrobiology become manageable when a systems approach was applied. I quickly wrote a book proposal to CRC, which was almost instantly approved. One reason, I suspect, is that there were no books on this topic available on the market! I started writing about applying the systems approach to the complex astrobiology problems I was familiar with, and which were amenable to a systems approach, and I experienced a rapid progress on this project.

However, when I considered the systems approach to the exploration of space and search for extraterrestrial life on Mars, I decided that the book would greatly benefit from the expertise of a space scientist. I was extremely fortunate to recruit as the co-author of the book the well-known space scientist, Benton C. Clark III, who likes to be called Ben. He agreed to fill in the blanks and enrich the book with his expertise. His knowledge of the systems extends to the robotic systems, such as Mars rovers. This, I figured, the readers of this book will positively love!

Vera Kolb
October 20, 2022

Acknowledgment

Thanks are expressed to Dr. Danny Kielty and other editorial and production staff from CRC, for their help in producing this book.

1 What Is Astrobiology?

1.1 DEFINITION OF ASTROBIOLOGY

Astrobiology is a scientific field that studies the origin, evolution, distribution, and future of life on Earth and in the universe (Des Marais et al., 2008; Des Marais, 2019, pp. 15–26; Kolb, 2019a, 2019b, pp. 3–13; Kolb and Clark, 2020). Astrobiology seeks to solve complex problems of the origin of life on Earth, the nature of the chemical evolution that led to life, characteristics of the transition between abiotic (not alive) chemical systems to biotic (alive) ones, and types of energy processes and chemical energy storage systems which emerged and ultimately led to the out-of-equilibrium thermodynamic status of the biotic systems, among other topics. Table 1.1 presents astrobiology goals in more detail.

TABLE 1.1

Astrobiology Goals, Themes, and Research Topics

1. *Understanding the origins, evolution, distribution, and the future of life in the universe*
   - Elucidate nonbiological origins of organic compounds in space and in planetary environments
   - Identify abiotic sources of organic compounds
   - Study the role of the environment in the production of organic molecules, and the influence of the environment on the stability and accumulation of organic molecules
   - Investigate the origins of organic compounds in space, their diversity and complexity
2. *Researching the origins and evolution of planetary systems*
   - Search for habitable environments, including those of the exoplanetary systems
3. *Understand the origins of life by investigating:*
   - Chemical pathways which can lead to the present-day biological constituents (such as DNA, RNA, proteins, and lipids)
   - Synthesis and function of macromolecules which are relevant to the origins of life
   - Information transmission
   - Chemical evolution that led to life
   - Self-organization processes in molecular evolution
   - The role of the environment in the chemical processes
4. *Early life and increasing complexity*
5. *Life and habitability*
6. *Co-evolution of life and the physical environment*
7. *Biosignatures for life detection:*
   - Distinguish life from nonlife
   - Follow the energy: identify energy sources, redox couples, and photoreactions
   - Follow biosignatures with time
8. *Search for extraterrestrial life*
9. *Spaceflight missions*

Source: A selection adapted and modified from Des Marais (2019), pp. 15–26.

DOI: 10.1201/9781003225874-1

1.2   IS ASTROBIOLOGY A DISCIPLINE?

Astrobiology is a field of study, rather than a discipline. It draws upon its foundational disciplines, such as physics, chemistry, biology, geology, and astronomy. Astrobiology also relies on various subdisciplines and research areas, such as atmospheric science, oceanography, evolutionary science, paleontology, planetary science, biochemistry, molecular biology, microbiology, ecology, and the history of science and philosophy. In addition, astrobiology is increasingly linked to various space missions, some of which have a direct goal to search for evidence for extraterrestrial life. Table 1.2 lists the foundational disciplines and subdisciplines of astrobiology.

1.3   ASTROBIOLOGY AS A FIELD OF STUDY

Astrobiology is a multidisciplinary, interdisciplinary, and transdisciplinary field of study (Kolb, 2019b, pp. 3–13). It is multidisciplinary since it takes simultaneously the perspective of several of its foundational disciplines. The interdisciplinary nature of

---

TABLE 1.2

Foundational Disciplines and Subdisciplines of Astrobiology, in Alphabetical Order

Analytical and instrumental chemistry

Astronomy

Atmospheric science

Biochemistry

Biology

Chemical evolution

Chemistry

Cosmic evolution

Cosmology

Cultural evolution

Ecology

Engineering

Evolutionary biology

Geology

Microbiology

Molecular biology

Oceanography

Paleontology

Planetary science

Philosophy

Physics

Robotics

Space engineering

Source: Kolb (2019b), pp. 3–13.

---

astrobiology is due to the extensive integration of these disciplines, their overlap, and the requirement for their synchronous examination. Transdisciplinary means that a synergy of views of the foundational disciplines is achieved and the conclusions are reached which are above and beyond those of any discipline, single or integrated with others (Repko, 2012).

Astrobiology is informed by its foundational disciplines, which include their sub-disciplines, and related research areas. All these need to interact among themselves in such a way as to create new competences which are not achieved by any individual part. To achieve this goal, astrobiologists need to adopt "systems thinking" in which astrobiology is a systems science, and various disciplines and subdisciplines are its components. Importantly, the National Academy of Sciences' most recent report recommends a systems approach to solving complex astrobiology problems. This report points out that many other modern fields now employ the systems approach, with great success, and thus astrobiology should do the same (National Academy of Sciences, 2019). The transdisciplinary feature of astrobiology makes it especially amenable to systems analysis, in which the system is more than a sum of its parts.

We offer here a brief introduction to the systems approach for the purpose of understanding this chapter. More about this approach will be presented in Chapter 3.

A system is composed of parts that interact with each other in such a way to produce a new, emerging property that is a characteristic of the system which is not contained in any of its individual parts. Systems are typically characterized by networks, feedback loops, and other complex interactions, between its parts. Importantly, the parts of a system may be quite complex in their own right, and in the context of the system need to be treated as subsystems.

As one learns the scope of the systems approach, one realizes the huge capabilities it has for the advancement of the field of astrobiology. These capabilities will be gradually introduced, presented, and elaborated on throughout this book.

Some elements of the general systems approach/analysis may not be directly applicable to some astrobiology goals and need to be modified. One example is that of an engineering system, which must have a pre-stated purpose, with a predetermined set of detailed requirements. Such a purpose is not presumed for some astrobiology goals, such as the chemical evolution or evolution of life in general. It would be very difficult to define ahead of time the outcome of the evolutionary process. The way around this problem is to look at the "purpose" as it is actualized in the functioning of the system.

2 Examples of Complex Problems in Astrobiology

One of the most difficult complex problems in astrobiology is how life originated on the Earth. This problem has many subsidiary problems, such as how the primordial metabolism, genetic system, and compartmentalization took place. These, in turn, generate related complex problems, such as those of prebiotic chemistry, by which the biologically relevant compounds were synthesized. For these reactions to occur, chemical energy sources need to be identified. Another complex problem is how the out-of-equilibrium thermodynamic requirement for an alive system evolved or emerged from the primitive protocells in conjunction with their environment, and how this requirement was maintained to keep life going.

There are other complex astrobiology problems such as an interplay between prebiotic chemistry and the geological conditions on the early Earth. Furthermore, the early Earth was bombarded by meteorites and asteroids, which caused an influx of chemicals and localized thermal energy, but also a change of habitability and suitability for an origin of life at various locales on the Earth.

There are also problems with classification of various biologically related entities, such as viruses, which are mysterious since they exhibit some but not all the properties of alive systems. Thus, it is difficult to decide if viruses are alive or not. But before we can decide what is alive and what is not, we need to agree on what life is. This seems to be obvious, but upon close critical examination, it turns out that it is not, and, thus, defining life is also a complex problem. We shall cover the definition of life in Chapter 4, and the case of viruses in Chapter 10.

Another difficult complex problem in astrobiology is that of determining the nature of the putative extraterrestrial (ET) life. Is it the same as ours, or is it similar, or perhaps radically different? How can we decide on this when we have not found the ET life? Knowledge gaps in this area are huge, and at this time we cannot overcome them. Still, we can at least develop an approach that will be the beginning of solving this problem. We need to first define life as it is on the Earth. Then we need to propose the nature of ET life, which will be based on the essential characteristics of life on Earth. Based on these, we can formulate the ET life's subsystems that are functionally equivalent to our life, but not necessarily the same. These subsystems will also need to accommodate different conditions under which ET life evolved and persisted. More subsystems may need to be added, especially in terms of energy, such as UV light, cosmic radiation, and solar emissions, as some examples. The space missions and the remote sensing within our Solar System, as well as the astronomical observations of exoplanets, are rapidly generating data, which allow the systems analysis in these areas, and which were out of reach not so long ago.

DOI: 10.1201/9781003225874-2

In Chapter 11, we examine the possibility of interplanetary transfer of life within our Solar System, specifically from Mars to Earth. We show that such a successful transfer may occur only at specific times in the history of these planetary subsystems.

Some complex astrobiology problems may be linked to other complex problems, which creates a perception of overwhelming "mega" problems. The first instinct is to go back to the reductionistic approach, which limits the number of components and their interactions, and then study them in isolation. However, the systems approach may be more productive for solving such problems. As one example, instead of examining separately the metabolism-first, the genetic system-first, and the compartmentalization-first systems, it could be more fruitful to study them as subsystems which co-evolved among themselves and with the environment. Such co-evolution would eventually lead to the primitive protocells. Inclusion of the environment in this process is necessary, especially to explain the out-of-the-equilibrium thermodynamic conditions which are characteristic of life and had to be established at some level of development in the protocells.

The systems approach is also helpful for studying prebiotic chemical systems. To make the laboratory experiments and models for these more adequate, one needs to consider also chemicals that were periodically delivered to the early Earth, by asteroids, meteorites, comets, and interstellar dust. Such exogenous delivery of chemicals was intermittent, and not sustainable over a long period of time, which complicates the modeling of the prebiotic chemical systems. We cover prebiotic chemistry mostly in Chapter 5.

3 An Overview of the Systems Approach to the Understanding and Solving the Complex Problems Relevant to Astrobiology

3.1 INTRODUCTION

Scientists in general and astrobiologists in particular need to become more familiar with the systems approach to understanding and solving complex problems. This method involves systems thinking and analysis. It supplements the reductionist method for studying and solving scientific problems. The latter breaks up the problem into the smaller parts and then studies these parts separately. This approach is commonly used in science and is valuable since it informs us about the nature and the properties of the parts of the system. However, the understanding of the parts does not necessarily help us understand how the system as a whole functions, especially if it is complex. Systems approach looks at the whole from the point of view of how the parts within the whole interact with each other to give the whole new or emergent properties, which enable its various functions. These new properties are not revealed by the individual parts. The interaction of the parts, including their organization and formation of networks, is the key to understanding the functioning of the whole and of its emergent properties.

As we have stated earlier, the systems approach is not sufficiently utilized by astrobiologists, although some applications have been reported. We give an example in Section 3.2. In Section 3.3, we provide a simple example of the need for the systems approach in astrobiology. Then, to facilitate and encourage the use of the systems approach in astrobiology, we explain this approach in general in Section 3.4.

3.2 AN EXAMPLE OF PREVIOUS APPLICATIONS OF THE SYSTEMS APPROACH TO ASTROBIOLOGY

In 2013, Chela-Flores published an influential paper titled "From systems chemistry to systems astrobiology: life in the universe as an emergent phenomenon" (Chela-Flores, 2013). The author considered systems astrobiology and the emergence of life in

DOI: 10.1201/9781003225874-3

the universe in general, not only on the Earth. He outlined the following stages of the systems approach to astrobiology, some of which could be experimentally validated:

1. The development of the theory, in which the universe is treated as a complex system with evolutionary convergence.
2. The computational modeling of astronomical and spectroscopic data.
3. The development of a testable hypothesis, in which Earth-like exoplanets are studied for the presence of biogenic gases.
4. The consideration of the entire universe as a complex system.
5. The experimental validation in which data are gathered from searches for exoplanets by various remote sensing means.

Chela-Flores's work (2013) could be used as a roadmap for multiple applications of systems approach to astrobiology. His goals, however, are very broad, and many are not fully achievable in the near future.

Our approach is not so all-encompassing. We focus mainly on the origin of life on Earth and possibly at some other places in our Solar System. Regarding the exoplanets, the distance between planetary systems with different stars is so large that interactions are minimal. Thus, those systems are very tenuous, although worthy of study. We cover later in Section 11.2 the study of lithopanspermia between Earth and Mars, which can be a model for exchange of life between other proximal planets.

3.3   AN EXAMPLE OF THE NEED FOR A SYSTEMS APPROACH IN ASTROBIOLOGY

We consider the cell, which is the system central for life, and thus critical for understanding of astrobiology. It contains a great number of chemicals, whose structures are given together with their functions in biochemistry books (e.g., Voet and Voet, 2011). If we break up the cell and extract all the chemical parts from the cell, such as amino acids, sugars, nucleic acids, various proteins, and inorganic ions, and then mix them all back together, these parts will constitute a heap of chemicals but this heap will not be alive.

In Figures 3.1–3.4, we give examples of both simple and more complex chemicals that are found in cells.

The above figures show that some chemical parts of the cell are chemically simpler than others. There are more complex chemical parts that we have not shown. They are available, e.g., in Voet and Voet (2011).

A heap of these chemical parts does not constitute an alive system, since it lacks critical interactions, organization, and other complex features that enable the system to exhibit properties of life. When we break up the cell and extract its chemical parts, we destroy the complex features that enable cells to function as an alive system. A mixture of extracted parts will not become alive on its own. It will not become an alive cell. It turns out that the knowledge of the details of its parts is not sufficient to explain how the cell functions as a unit of life.

How can we solve this problem? The first instinct is to look more into the details of the individual parts and seek an answer there. While no knowledge should be

Examples of chemical parts of a cell

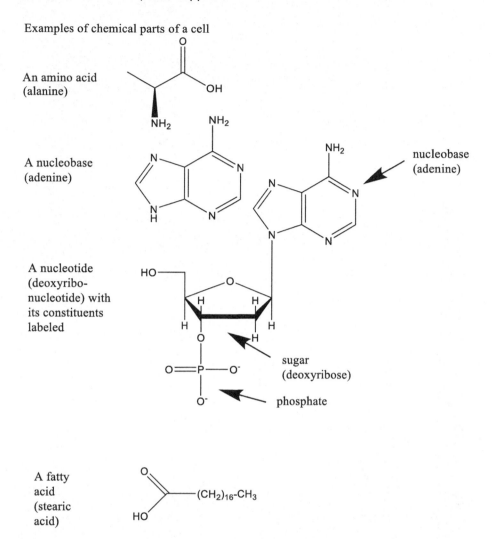

FIGURE 3.1   Examples of chemical parts of a cell. (Substantially modified from Voet and Voet, 2011.)

discarded or classified as unimportant, by pursuing the study of the individual parts we may end up knowing more and more about an individual part, but we still will not be able to understand the way the system operates. Is there a way to look at the parts of the system to avoid this problem?

The answer is yes. When we focus on the parts of the system, such a focus should be within the context of the whole system. Thus, one focuses on the part, but does not lose sight of the system. This enables us to observe the part not as an isolated entity, but as a connected, dynamic, interactive one. Eventually, we will reach an understanding of a system as a whole and the functioning of the parts within it, as the thinking process moves more from the parts-oriented to the whole-oriented.

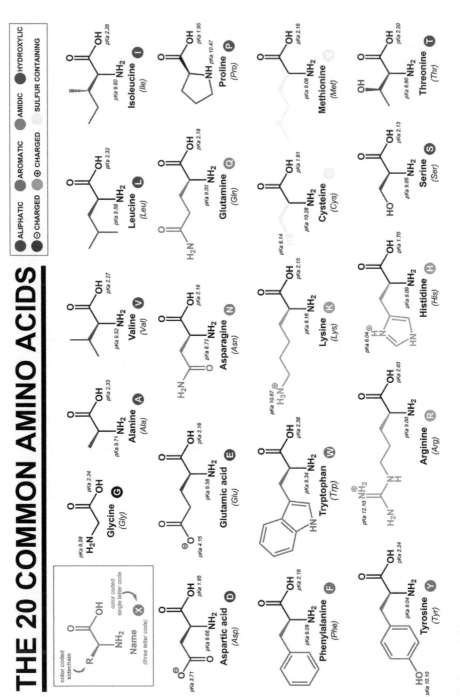

FIGURE 3.2 Structures of the common 20 amino acids. (With permission from Shutterstock.)

Biology ● ● ●
Adenosine Triphosphate (ATP)

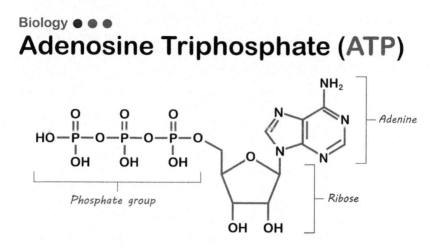

● **Energy-carrier in all of living things**

● **Consist of nitrogenous base (adenine), sugar (ribose) and phosphate group**

FIGURE 3.3   The structure of ATP, the energy unit for driving chemical processes in the cell. (With permission from Shutterstock.)

Structure of RNA and DNA

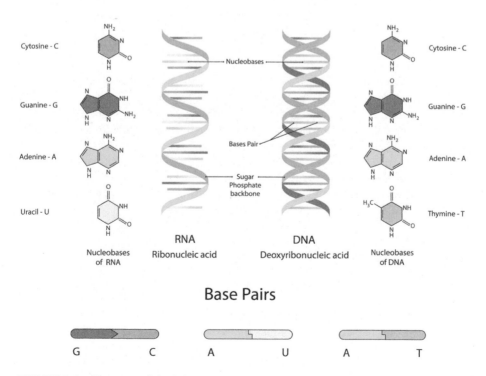

FIGURE 3.4   Structure of the informational molecules, RNA and DNA. (With permission from Shutterstock.)

3.4 DEFINITION AND CHARACTERISTICS OF A SYSTEM, SYSTEMS THINKING, AND SYSTEMS ANALYSIS METHOD

3.4.1 OUR FOCUS AND A BRIEF BACKGROUND

Our overview of the systems thinking/analysis method does not provide specifics for all the fields in which this method is commonly employed, such as engineering, economics, management, and the computer sciences, and it does not include all the details that are commonly found in the general books and articles about this method. Instead, we describe the main characteristics of the systems in general, and the critical aspects and details of the systems/thinking analysis which are relevant to astrobiology. We draw also upon the relevant applications in fields such as education, chemistry, biology, and geology.

Much can be learned from the books on the broad subject of systems (e.g., von Bertalanffy, 1993; Capra, 1996; Capra and Luisi, 2014; Lovelock, 1995; Meadows, 2008; Kasser, 2013; O'Connor and McDermott, 1997; Systems Thinking and Practice, 2016) and articles, which we cite as we discuss them. We extract the main points that are helpful for studying astrobiology systems.

Systems, systems thinking, and systems analysis are often covered in the same literature resources. These topics are often linked. A paper that discusses the systems thinking usually also addresses the characteristics of the system and justifies a need for the systems thinking which is necessary for the understanding of the functioning of the system. To gain this understanding, one needs to perform a "systems analysis." We present an overview of these topics below. There is some overlap between the definitions and descriptions as given by different authors. To improve readability and to properly credit the individual authors, we did not attempt to eliminate such an overlap.

A system is a group of parts that interact to form a unified whole. The system is said to be more than the sum of its parts, since it exhibits novel or emergent properties, which are not found in the individual parts. A system is characterized by its boundaries, organization, and purpose which is expressed in its functioning. The system is surrounded and influenced by its environment and may be subject to the arrow of time (as in an evolutionary system).

The parts of a system may differ in their complexity. Simpler parts are often referred to just as the parts or the components of the system. The complex parts/components may constitute their own system, which then needs to be considered as a *subsystem* of the larger system to which it belongs.

3.4.2 IN-DEPTH COVERAGE

We first present selected definitions of systems thinking from the paper "A definition of systems thinking: a systems approach," by Arnold and Wade who summarized and critically assessed this topic (Arnold and Wade, 2015, and the references cited therein). We pick and choose definitions that we believe are most useful to astrobiology. In addition, we shorten and paraphrase the original definitions to bring out their relevance to astrobiology. These definitions are presented in Table 3.1. The reader is

TABLE 3.1

Definitions of Systems Thinking (Modified from Arnold and Wade, 2015) Which Are Particularly Useful to Astrobiology

1. System thinking is the ability to see both the forest and the trees; one eye on each (based on Barry Richmond's definition)
2. Systems thinking is the ability to include multiple perspectives and operate within a space where the boundary of the system may be fuzzy (based on Squires et al's definition)
3. Systems thinking enables seeing wholes, interrelationships between parts, and patterns of change rather than static snapshots (based on Peter Senge's definition)
4. Systems thinking is the ability to recognize interconnections, to identify feedback, and to understand dynamic behavior (based on Hopper and Stave's definition)
5. Systems thinking is the ability to understand how the behavior of a system arises from the interaction of its parts over time, namely, to assess dynamic complexity; to discover feedback processes; and to identify nonlinearity (based on Sweeney and Sterman's definition)

referred to the original paper for the details and the original references. We do not cite the latter, due to space constraints.

The definitions from Table 3.1 capture the essence of what systems thinking is. We suggest that the readers revisit this table often.

We next present selected excerpts from the paper "Navigating complexity using systems thinking in chemistry, with implication for chemistry education" (Constable et al., 2016), and the references therein), which address the main points of the paper, including the purpose of the system. They are shown below, followed by our comments.

A system comprises a set of elements working together to form a complex whole that produces a function (also see Meadows, 2008).

This definition is quite valuable for astrobiology since it takes away the need to specify the purpose of the system, which is common in the fields such as engineering, but is difficult to ascertain in astrobiology. This is particularly important for astrobiological goals of studying life and its origins on Earth. We can then say that the "purpose" of life is to function. The life's function is in accordance with its key properties, such as metabolism and reproduction with inheritance. Life also evolves.

Complexity is a hallmark of systems.

This is important since it warns us to resist reductionist ways of trying to dissect and over-simplify a system. Instead, we need to learn how to deal with the complexity of the system as it is.

Complexity differs from system to system. The nature of the complexities associated with each discipline…differ. Moreover, complexities between subsystems within a discipline also differ….

Here we learn that the understanding of one complex system or subsystem does not guarantee the understanding of a different one. However, the systems thinking ability will help us conquer new cases.

We next summarize the sections from Constable et al. (2016) on "Process for characterizing the system" and "Building systems complexity: adding boundaries and loops" and point out their relevance to astrobiology.

Characterization of a system starts with considerations such as the choice of scale, boundaries, hierarchies, and constraints. We must also examine the way in which the system's relationships change over time and distance, the nature of interfaces between the system and subsystems, the feedback loops interactions, and the emergence of system's functions. Further, we must pay attention to what happens beyond the boundaries of the system under study, since the system may expand in the future, including the origin of the material that comes into the system, and the source of energy provided to the system. All of these are critical for astrobiology, as we shall see from the ample examples which will be given in the later chapters.

Constable et al. (2016) also discuss the system's purpose. They point out that systems are often described as having a purpose, but that this can be confusing in the context of chemistry. We have pointed out earlier in this section that it is also confusing in the astrobiology context. However, as we stated above, we avoid this confusion by adopting the definition that the purpose of a system is to produce a function (Meadows, 2008). We thus focus on the system's function, and do not ponder about its purpose above and beyond this function.

A particularly important section of Constable et al. (2016) paper addresses the character of emergence in chemistry systems. We cite: "Broadly, the concept of emergence in chemical systems refers to phenomena in which the structures, properties, or behavior of multicomponent systems exceed those predicted from knowledge of the individual components." The authors also briefly address systems chemistry and its studies of tendencies of some molecules toward self-assembly that creates membrane-like structures or other biologically important constructs. These are strongly related to astrobiology goals. We shall cover much more about systems chemistry in the forthcoming chapters in which we address prebiotic chemistry and the emergence of life from abiotic systems.

We draw upon another paper on systems thinking in chemistry, "ChEMIST table: A tool for designing or modifying instructions for the systems thinking approach in chemistry education" (York and Orgill, 2020, and the references cited therein). We again extract the material which we consider especially valuable for astrobiology.

In the section of the paper "Need for an operational definition of systems thinking in chemistry education," the authors point out that without an operational definition or consensus for what systems thinking is, it is possible that this term could mean many different things to different people. They further cite research results that suggest that "systems thinking is not a 'natural' way for humans to think." Especially difficult is to think about systems as consisting of interacting dynamic components. The authors propose an operational definition of systems thinking in the context of chemistry education and point out that such a definition is not universally acceptable for all the fields. Table 3.2 summarizes their proposal. The reader will notice a partial overlap with the definitions of systems thinking from Table 3.1. However, a different wording of the same item is pedagogically useful.

We comment on the last item of Table 3.2. For an application to astrobiology, we need to be persistent in the search for emergent behavior, and not quit the search

TABLE 3.2

Operational Definition of Systems Thinking

1. The system is more than just a collection of parts
2. Components of the system are organized in a particular manner, and they interact with each other and with the environment
3. The system contains both positive (or reinforcing) and negative (or balancing) feedback loops
4. Properties of the system are dynamic (i.e., vary with time)
5. Nonlinear relationships between some variables and the system's behavior may exist
6. System's boundaries need to be established
7. The pattern of the system's behavior needs to be observed over time
8. Emergent behavior needs to be identified

Source: Adopted from York and Orgill (2020).

TABLE 3.3

Selection from Richmond's Systems Thinking Skills

1. "Dynamic thinking" is applied when the system's behavior changes over time; it may enable a glimpse into the future behavior of the system
2. "Forest thinking" examines the system's behavior as a whole instead of focusing just on the parts of the system; the latter is termed the "Tree-by-tree" thinking model

   "Closed-loop thinking" considers that variable 1 may affect variable 2, and vice versa, just like in the familiar case of metabolic feedback loops

Source: Adopted from Orgill et al. (2019), and the references cited therein.

at the first identification of such a behavior. There may be more than one emergent behavior.

We examine next another chemically oriented paper, "Introduction to systems thinking for the chemistry education community," by Orgill et al. (2019, and the references cited therein). These authors bring up insights into the systems thinking which are useful for astrobiology. They point out that systems may exist on different scales, such as microscopic or macroscopic. The boundary conditions for a given system are established by the observer. These authors discuss in detail Richmond's seven systems thinking skills, from which we select following material shown in Table 3.3. We do not re-cite their references, because of space constraints.

Importantly, these authors offer an especially useful "Hierarchical pyramid model" for systems thinking. This model builds up from the base of the pyramid in eight levels, where the base is level 1, and the highest and narrowest level is 8. We somewhat modify and supplement this model to suit astrobiology, as presented in Table 3.4.

The entries from Tables 3.3 and 3.4 are quite specific and can be used as systems thinking tools. We recommend that the reader re-visits them on a regular basis.

TABLE 3.4
The Pyramid Model for Systems Thinking

1. Level 1 is at the base of the pyramid; it focuses on identification of systems components
2. Level 2 has a goal of identifying relationships between the systems components
3. Level 3 concentrates on identifying dynamic relationships within the system
4. Level 4 targets the organization of the system's components and processes
5. Level 5 is the identification of feedback loops and various reaction cycles
6. Level 6 is the examination of behavior between related systems
7. Level 7 has a goal of understanding the features of the system which are "hidden" from view, namely not explicit
8. Level 8 involves thinking in retrospect, present, and the future of the system

Source: Modified from Orgill et al. (2019).

TABLE 3.5
Selected Complex Properties of the Systems

1. Resilience
   It is the tendency of a system to remain in the stable condition or return to it from an unstable state. This tendency is the result of a change in forces that act upon the system. Resilience is limited in scope and is related to the magnitude or type of the force which act upon the system.
2. A tipping point
   It is a threshold at which the complex system cannot return to the starting point.
3. Matter and/or energy exchanges
   These occur between the interacting parts within the system and its environment.
4. Transfer of energy or matter
   Such transfer may occur in a complex system. If the system is open, the transfer of energy or matter across the boundary occurs; if the system is closed, such transfer does not occur.
5. Positive feedback
   Such feedback amplifies the rate of change in the system.
6. Negative feedback
   It is opposite of the positive feedback, and thus it diminishes the rate of change in the system.
7. Nonlinear change over time
   Such change means that the magnitude of the response is not linearly proportional to the magnitude of the input.

Source: Based on Bar-Yam (2002); https://en.wikipedia.org/wiki/Complex_system.

Next, in Table 3.5, we show examples of selected complex properties of the systems.

Figures 3.5 and 3.6 illustrate tipping point, respectively, in a manner that is easy to understand.

In conclusion of this section, we have presented various definitions of systems, systems thinking, and systems analysis in different fields, and have chosen those that are most promising for use in astrobiology.

FIGURE 3.5   Resilience. (With permission from Shutterstock.)

FIGURE 3.6   Tipping point. (With permission from Shutterstock.) It is intuitive that a small change in the applied force on the system can drive it to the point of no return.

4 Application of Systems Analysis to Defining Life

4.1 INTRODUCTION AND AN OVERVIEW OF THE PROBLEM

Defining life is important for astrobiology for several fundamental reasons (e.g., Kolb, 2019c, pp. 57–64, and the references therein). Significantly, we need to define life in order to study its origin and evolution on the Earth, which are the principal objectives of astrobiology, so that we know what we are trying to study. Another key objective of astrobiology is the search for the putative extraterrestrial (ET) life. To anticipate what we are searching for, we need to have a working definition of ET life. Such a definition is usually modeled by what we know so far about life, which is life on the Earth, since we have not yet discovered the ET life. Another important goal of astrobiology is to try to understand the origin of life by creating artificial life. This would also require a definition of life to guide these efforts. Many specific aspects of astrobiology also require a definition of life. For example, the study of the abiotic-to-biotic ("a-2-b") transition, during which the matter which is not alive transitions to life, also needs a clear definition of life. Likewise, the study and the understanding of the critical characteristics of life, such as metabolism, information transfer, reproduction, and compartmentalization, also need the definition of life. One can come up with more and more such examples to justify the need for defining life. This endeavor is notoriously difficult. There are over 100 definitions proposed, but despite much discussion, there is no single definition that satisfies all the researchers. An extensive list of the definitions of life can be found in the literature (e.g., Pályi et al., 2002; Popa, 2004, 2015; Kolb, 2019c, pp. 57–64, and the references cited therein). We give a selection of the proposed definitions of life which exemplify different approaches to defining life. Some definitions of life use the systems approach to some extent and at various levels. These may include the term system or refer to some of the system's properties, such as the need of the parts within the system to interact among themselves to create networks and various levels of organization, from which life emerges.

In Section 4.2, we give a short background on definitions in general and provide selected examples of the definitions of life. The order in which we give definitions is not necessarily chronological; instead, we mix and sometimes match definitions to provide a variety to keep the readers' interest. Some old definitions are remarkably similar to the contemporary ones. We provide over 40 definitions but encourage the interested reader to explore all the definitions from the sources we cite.

DOI: 10.1201/9781003225874-4

4.2 BACKGROUND ON DEFINITIONS IN GENERAL AND EXAMPLES OF DEFINITIONS OF LIFE

We start with a brief background on definitions in general. Different types of definitions exist, such as lexical, stipulative, and operational definitions (Churchill, 1986, 1990; Gupta, 2015). A lexical definition is a dictionary definition. It gives the meaning of a term as it is currently used. A stipulative definition gives a new specific meaning to the term. It is a proposal to use the term in a certain way, typically for purposes of discussion in a specific context. An operational definition typically states the meaning of a term based on its observable properties which can be measured (Churchill, 1986, 1990). Each of these types of definitions is valuable for astrobiology. The lexical definition guides us about the common use of the term "life" as it is understood and used by the public. When astrobiologists propose a definition of life, they should be aware of the common knowledge about life that the general public is starting from. A stipulative definition is quite important to astrobiologists, since they can specify how they view the definition of life in the context of the problem they study. Examples may include defining life at its very beginning, or during its evolution that led to complex species. An operational definition is critical for the search for life in our Solar System or elsewhere in the universe, since it starts with some parameters that can be observed, which give us confidence that the definition of life is not completely abstract of even arbitrary. We provide below some examples of each type of definition of life.

An example of a dictionary (lexical) definition of life:

1. "Life 1a: the quality that distinguishes a vital and functional being from a dead body; b: a principle or force that is considered to underlie the distinctive quality of animate beings (compare vitalism); c: an organismic state characterized by capacity for metabolism, growth, reaction to stimuli, and reproduction ... 5 a: the period from birth to death" (Merriam-Webster's Collegiate® Dictionary, Tenth Edition, 1993). The more recent online version of this dictionary, https://www.merriam-webster.com/dictionary/life, visited on 9/28/2022, states: "1a: the quality that distinguishes a vital and functional being from a dead body; 1b: a principle or force that is considered to underlie the distinctive quality of animate beings; 1c: an organismic state characterized by capacity for metabolism, growth, reaction to stimuli, and reproduction........5a. the period from birth to death." This definition, which is actually a collection of definitions, has not changed in almost 30 years, other than to delete "compare vitalism" in the latest edition, possibly since vitalism is a discredited idea. However, the definition still keeps "vital," "animate," and "force," all of which are associated with vitalism.

   A case of a stipulative definition of life:

2. "Life is what the scientific establishment (probably after some healthy disagreement) will accept as life" (Friedman, 2002, paraphrasing Theodosius Dobzhansky; source Popa, 2004). This definition may be an extreme case of a stipulative definition.

   An illustration of an operational definition:

3. "Any definition of life that is useful must be measurable. We must define life in terms that can be turned into measurables and then turn these into a strategy that can be used to search for life. So, what are these? (1) structures; (2) chemistry; (3) replication with fidelity; and (4) evolution" (Nealson, 2002; source Popa, 2004). Items 1–3 of this definition survive the test of time. They are very practical and achievable. Item 4 represents a partial knowledge gap, since we cannot completely elucidate evolution, which is a historical process. This definition does not specifically address prebiotic evolution that led to life. If considered under "evolution" in part 4, then the knowledge gap of 4 becomes even larger.

Some definitions of life draw upon the history of ideas on this subject and view life as animation (Aristotle), as mechanism (Descartes), and as organization (Kant) (Gayon, 2010). Approaches such as holism, reductionism, and dialectical materialism also influence the process of defining life (Popa, 2010, 2015). For example, dialectical materialism considers life as a fully anticipated phenomenon, which is the result of the laws of nature (Popa, 2010). Such an approach is taken in the following definition of life:

4. "Life is a new quality brought upon an organic chemical system by a dialectic change resulting from an increase in quantity of complexity of the system. This new quality is characterized by the ability of temporal self-maintenance and self-preservation" (Kolb, 2002; source Popa, 2004; Kolb, 2019c, pp. 57–64; Zimmer, 2021, pp. 270–271).

Another definition that views life as an anticipated phenomenon is:

5. "Life is an expected, collectively self-organized property of catalytic polymers" (Kauffmann, 1993; source Popa, 2004).

Some definitions of life are expanded lexical definitions that include additional terms to reflect the needs of the specialists. They often focus on key biochemical components of life, which are then tied to defining life. Life is often defined by a list of properties that are common to all life. Such a list typically includes chemical components, compartments to enclose them, metabolism, reproduction, and thermodynamic properties (Gayon, 2010). Many definitions focus on self-replication/reproduction and mutation as essential for life. The terms "self-replication" and "self-reproduction" are often used synonymously, although some investigators point out that self-replication means making an exact replica such as in the case of macromolecular replication, while self-reproduction means making a similar copy, and in the case of cell reproduction (Luisi, 2006, 2016). Let us point out that "self-replication" is a term that requires some clarification. As just one example, an organism that undergoes sexual reproduction cannot generally self-reproduce. However, there are individual life forms and entities that cannot reproduce, such as sterile organisms, but are obviously alive (the classic example being the mule). The requirement for reproduction is clearer when placed in the context of species, rather than an individual. This problem has been addressed in the literature (e.g., Kolb, 2007, 2019c, pp. 57–64). For viruses that cannot reproduce on their own, a question if they are alive or not persists. In Chapter 10 we shall suggest a solution to this problem by systems analysis.

Examples of definitions of life that focus on capturing the perceived essential characteristics or functions of life:

6. "The characteristics that distinguish most living things from non-living things include a precise kind of organization; a variety of chemical reactions that we term metabolism; the ability to maintain an appropriate internal environment, even when the external environment changes (a process referred to as homeostasis); and movement, responsiveness, growth, reproduction, and adaptation to environmental change" (Vilee et al., 1989; source Popa, 2004). This definition would benefit from defining these characteristics more fully. For example, what is exactly a "precise" kind of organization? Also, the characteristics are given as separate items, without showing how they are connected.

7. "Any living system must comprise four distinct functions: (1) increase of complexity; (2) directing the trends of increased complexity; (3) preserving complexity; and (4) recruiting and extracting the free energy needed to drive the three preceding motions" (Anbar, 2002; source Popa, 2004). This definition depends on the understanding of what complexity is. This is a common problem in various other definitions which we list below. It is unclear how complexity can be preserved (3) when it should be increased (1). Furthermore, there are many examples of life forms that have evolved to less complexity, such as most parasitic organisms.

8. "Life is synonymous with the possession of genetic properties, that is, the capacities for self-replication and mutation" (Horowitz, 2002; source Popa, 2004).

9. "Life is a chemical system capable of transferring its molecular information independently (self-reproduction) and also capable of making some accidental errors to allow the system to evolve (evolution)" (Brack, 2002; source Popa, 2004).

10. "Any system capable of replication and mutation is alive" (Oparin, 1961; source Popa, 2004).

Examples of definitions that include the term "life forms":

11. "Life is a chemical phenomenon that occurs in space and time as a succession of life forms that combined have a potential to metabolize, reproduce, interact with the environment, including other life forms, and are the subject of natural selection" (Kolb and Liesch, 2008).

12. "We propose that life of an organism is the sum of its life forms over a period of time. We set the integral of time from the birth of the organism to its death. Reproduction, particularly by fission, represents an identity dilemma, but it can be resolved by Gallois' occasional identities theory" (Kolb, 2010).

A definition of life that is accepted by most researchers is:

13. "Life is a self-sustained chemical system capable of undergoing Darwinian evolution" (NASA's working definition of life; Joyce, 1994, 2002; source Popa, 2004). This definition would benefit from clarifying what "self-sustaining" is. It also assumes that a chemical system can undergo Darwinian evolution.

We have only scratched the surface of the problem of defining life. Many more aspects of this problem are reviewed (Kolb, 2019c, pp. 57–64, and the references therein), such as the pros and cons of minimal vs. expanded definition of life; application of Aristotelian principles to defining life; borderline cases of life such as viruses; and philosophical problem of identity and its application to defining life (Kolb, 2005, 2007, 2010, 2019c, pp. 57–64).

Some of the above-listed definitions use the term system and/or some of the critical properties of a system. We list additional such definitions.

14. "Living things are peculiar aggregates of ordinary matter and of ordinary force which in their separate states do not possess the aggregates of qualities known as life" (Bastian, 1872; source Popa, 2004). This is an early definition that is valuable since it points out that separate states do not have the qualities of their aggregates. This is reminiscent of the isolated parts of the system not having properties of the system.

15. Life is "a hierarchical organization of open systems" (von Bertalanffy, 1993; source Popa, 2004). This is a valuable definition that emphasizes organization, and open system properties.

16. "A living system is an open system that is self-replicating, self-regulating, and feeds on energy from the environment" (Sattler, 1986; source Popa, 2004). This definition specifically includes energy from the environment that life needs.

17. "Living units are viewed as objects built of organic compounds, as dissipative structures (or at least dynamic low entropy systems significantly displaced from thermodynamic equilibrium)" (Prigogine, 1980; Prigogine and Stengers, 1984; reformulated by Korzenieweski, 2001; source Popa, 2004). This definition is thermodynamically based, and it considers dissipative structures.

18. "In order to be recognizable life must: 1. be a nonequilibrium chemical system; 2. contain organized polymer; 3. reproduce itself; 4. metabolize by itself; 5. be segregated from the environment" (Buick, 2002; source Popa, 2004). The nonequilibrium state of life as a system is brought up.

19. "Life is a near-equilibrium, tightly controlled, open, dissipative, complex system. Such a system can only work if its parts are 'designed' (by evolution) to push thermodynamics to its limits" (Hoffmann, 2012).

20. "Life is a population of functionally connected, local, non-linear, informationally-controlled chemical systems that are able to self-reproduce, to adapt, and to coevolve to higher levels of global functional complexity" (von Kiedrowski, 2002; source Popa, 2004). Life is defined as a chemical system. It is not clear what "global functional complexity" is.

21. "Living organisms are systems characterized by being highly integrated through the process of organization driven by molecular (and higher levels of) complementarity" (Root-Bernstein and Dillon, 1997; source Popa, 2004). This definition points out molecular complementarity.

22. "Life as a system capable of (1) self-organization, (2) self-replication, (3) evolution through mutation, (4) metabolism, and (5) concentrative encapsulation" (Arrhenius, 2002; source Popa, 2004). Life is defined as a system. Its main capabilities are listed.

23. "A terrestrial living entity is an ensemble of molecular-informational feedback-loop systems consisting of a plurality of organic molecules of various kinds, coupled spatially and functionally by means of template-and-sequence directed networks of catalyzed reactions and utilizing, interactively, energy and inorganic and organic molecules from the environment. A living entity is an uninterrupted succession of ensembles of feedback-loop systems evolved between the emergence time and the moment of observation" (Lahav, 2002; source Popa, 2004). This definition is comprehensive since it specifies the use of energy and its interaction with the environment. It points out the continuity of life once it emerged.

24. "… a proposed definition of *life*: an open, coherent spacetime structure kept far from thermodynamic equilibrium by a flow of energy through it – a carbon-based system operating in a water-based medium with higher forms metabolizing oxygen" (Chaisson, 2003). This definition points out spacetime (spatiotemporal) nature of life systems, as well as its far-from-equilibrium thermodynamic status.

25. "Life is a metabolic network within a boundary" (Maturana and Varela, 1980; reformulated by Luisi, 1993; source Popa, 2004). "All that is living must be based on autopoiesis, and if a system is discovered to be autopoietic, that system is defined as living; that is, it must correspond to the definition of minimal life" (Maturana and Varela, 1980). The term "autopoiesis" is the property of regulating composition and preserving boundaries. Maturana and Varela stated that the concept of autopoiesis is necessary and sufficient to characterize the organization of living systems. However, this characterization does not include information about the physicochemical structure of the system's components to supplement the abstract notion of self-organization.

26. "According to the systems view, the essential properties of an organism, or living system, are properties of the whole, which none of the parts have. They arise from the interactions and relationships between the parts. These properties are destroyed when the system is dissected …into isolated elements" (Capra, 1996, p. 29).

27. "Consider Gaia theory as an alternative to the conventional wisdom that sees the Earth as a dead planet made of inanimate rocks, ocean, and atmosphere, and merely inhabited by life. Consider it as a real system, comprising all of life and all of its environment tightly coupled so as to form a self-regulating entity" (Lovelock, 1995).

28. "Life is a far from equilibrium self-maintaining chemical system capable of processing, transforming and accumulating information acquired from the environment" (Vitas and Dobovišek, 2019).

29. "…the living state can be thought of as a new state of matter, *the replicative state of matter*, whose properties derive from the special kind of stability that characterizes replicating entities – DKS. That leads to the following working definition of life: *a self-sustaining kinetically stable dynamic reaction network derived from the replication reaction.* Each word in the definition imparts an important element to the definition. 'Self-sustaining' means

that the system must have an energy-gathering capability in order to satisfy the requirements of the overriding Second Law. The terms 'kinetically stable' and 'dynamic' describe the characteristics of that other stability kind, and the words 'network' and 'replication' are self-explanatory..." (Pross, 2016, pp. 163–164). Pross contrasts chemical stability of the regular chemistry, which leads to thermodynamic equilibrium, with the dynamic kinetic stability (DKS) which applies to replicating systems, which leads to an out-of-equilibrium system. The DKS is really persistence. "Overriding Second Law" simply means that one has to pump energy into the system to prevent it from moving to the equilibrium state, or, alternatively, to maintain its out-of-equilibrium state.

30. "Life$_{Terra}$ is a genome-containing, self-sustaining chemical dissipative system that maintains its localized level of organization at the expense of producing entropy in the environment; which has developed its numerous characteristics through pluripotential Darwinian evolution" (von Hegner, 2019).

   Miscellaneous other definitions:

31. "All life is chemistry" (Jan Baptista van Helmont, 1648, in Ridley, 1999, p. 15; also in Gillian K. Ferguson, "The Human Genome: Poems on the Book in Life," blog: thehumangenome.blogspot.com/2008/05/dna.html; Tuesday, 29th April 2008).

32. "At least some life is chemistry" (Friedrich Wohler, 1828; in Ridley, 1999, p. 15; also in Gillian K. Ferguson, "The Human Genome: Poems on the Book in Life," blog: thehumangenome.blogspot.com/2008/05/dna.html; Tuesday, 29th April 2008).

33. "… we can formulate a new definition of life: life is communicative interaction, which means life is primarily a social event. …. Social events are realized by communicative interactions on three complementary levels in parallel: cell communication, RNA communication, and virus communication" (Witzany, 2020). See also Witzany (2015).

34. "…life is a material phenomenon, grounded in chemistry and physics. Life designates a quality, or property, of certain complex dynamic systems that persist by channeling through themselves streams of matter, energy and information. They have the unique capacity to reproduce themselves indefinitely, and arise on a millennial time-scale by the interplay of variation and selection that underlies biological evolution" (Harold, 2001).

35. "…we combine the two individual words of 'life' and 'form' into a single, all-representative word encompassing the minimum reproductive set of organisms, the Lifeform (Lf). …A Lifeform is a single organism, or a collection of specialist organisms, whose ability to reproduce is enabled by a set of indispensable yet modifiable instructions embedded in the Lifeform…. An Organism is any physical entity produced by a Lifeform that can, in a suitable environment, affect the flow and/or conversion of energy to perform active functions guided by a subset of the Lifeform's instruction set…By definition, an organism is 'alive'" (Clark, 2019b, pp. 65–74). This is a generalized and universalized definition of life. It permits non-self-reproductive

entities, such as the mule, children, eunuchs, red blood cells, and viruses to be considered as alive.

36. "… life is (1) composed of bounded microenvironments, (2) capable of transforming energy and the environment to maintain a low-entropy state, and (3) capable of information encoding and transmission" (Schulze-Makuch and Irwin, 2004).

37. "I would argue that the only satisfactory definition of life …lies in the most critical property of genetic heredity: independent evolution. Life is the manifestation of a coherent collection of genes that are competent to replicate within the niche in which they evolve(d). Viruses fulfill this definition" (Bhella, 2016).

38. "Life represent life as we know it; it uses the specific disequilibria and classes of components of earthly life". "…dissipation of free energy is …the first necessary aspect of life….it must be accompanied…by autocatalysis, homeostasis and learning – to form a sufficient description of the living state" (Bartlett and Wong, 2020).

39. "…a living organism must possess, at least, the following three properties: (i) the ability to process and transmit heritable information to progeny (i.e., a genetic mechanism); the ability to capture energy and material resources, staying away from thermodynamic equilibrium (i.e., metabolic machinery); and (iii) the ability to keep its components together and distinguish itself from the environment (i.e., cell membrane)" (Ruiz-Mirazo et al., 2014). These authors also present life graphically in a figure in which three areas, metabolic machinery, genetic/template mechanisms, and membrane compartment, are shown as partially overlapping rectangles, where the area of overlap is life.

40. "…living system must be able to: 1. *manufacture* ifs own constituents from materials available in its surroundings. 2. extract *energy* from its environment and convert it into the various forms of work it must perform to stay alive; 3. *catalyze* the many chemical reactions required to support its activities; 4. *inform* its biosynthetic and other processes in such a way as to guarantee its own accurate reproduction; 5. *insulate* itself in such a way as to keep strict control over its exchanges with the outside; 6. *regulate* its activities so as to preserve its dynamic organization in the face of environmental variations; 7. *multiply*" (De Duve, 1991). The author calls these "The seven pillars of life." He states that "these seven properties are both necessary and sufficient for life as we know it to exist and persist." We note that evolution via mutations is not covered, since in the pillar 4, only accurate reproduction is given.

41. "Life is a self-organizing, self-regulating entropy-maximizing iterator composed of a hierarchy of stable dissipative structures" (Abramov et al., 2021). The authors further explain: "'Self-organizing' refers to retrieving order from bifurcation and deterministic chaos via 'self-regulating' coupled feedback; 'iterator' refers to feedback cycles, or iterations; 'entropy maximization' and 'dissipative structures' are grounded in well-established thermodynamic concepts; 'hierarchy' refers to lower-order building blocks producing higher-order building blocks; and 'stable' means that continuous external energy input is not required."

42. "The traits of living systems are the following: the ability they have to reproduce; the fact that they possess an identity; the fact that biological events should be considered in the context of a history; the fact that living systems are able to evolve by selection of alterations of their structure and self-organization......Another property of living systems is that their behavior is defined in the context of a time-arrow" (Ricard, 2010a, Part I).

From the given selection of various definitions of life, we can see that each definition is valuable in its own way, as it brings up some important features of life. Some definitions are quite abstract; some others are rich in details. And yet, none is completely satisfactory to all researchers.

We notice that some of the problems in defining life arise from the lack of a systems approach. We give the example of the dictionary definition 5a, according to which life is the period from birth to death. Let us remind ourselves that the dictionary definitions are the ones that are commonly accepted by the general public. We thus pose a question: What can possibly be wrong with definition 5a?

Let us take a human as an example. Both birth and death are the events that are reasonably well understood. Let us now focus on death. First, we acknowledge that humans have many bacteria in their GI (gastrointestinal) systems which co-exist with humans and have an intimate relationship. Namely, there is a human-bacteria system. It is easy to see that when a human being dies, the bacteria in it may live longer and possibly transfer to another host in which they will continue living. Thus, the death of all the parts of humans does not generally occur at the same time. Likewise, cerebral brain death may occur while the body continues to be metabolically active for years before the cardiovascular system ceases to operate.

Another example is that of viruses. In the previously proposed definitions of life, viruses have not been given the status of being alive since they cannot self-reproduce. (More about this topic will be presented in Chapter 10.) However, if we consider the virus and its host as a system, then the viruses can reproduce in the host and therefore become alive. If we consider the time of death, we again see that the parts of the system need not die at the same time. For example, the virus may attack its host causing its death, but the virus will be released from the host and will then seek another host. Alternatively, the host may kill the virus. The host will survive, and the virus will die.

Popa stated (2010):

We may never agree on a definition of life which will remain forever subject of a personal perspective. The measure of one's scientific maturity may actually be his/her latest definition of life and the acceptance that it cannot be ultimate.

Based on the above selection of definitions of life and Popa's statement, it appears that it is difficult and perhaps fruitless to attempt to contribute yet another definition of life, above and beyond what the others have already done.

We start by explaining why we think that it is a worthwhile endeavor.

We acknowledge that all definitions of life that have been proposed have something important to convey. We draw on these definitions as a background for developing our suggested new definition.

However, scientific definitions in general and the definition of life in particular may need to change over time to accommodate progress in science. Definitions of life should be flexible enough to accommodate such progress. Thus, instead of aiming to formulate an absolute, unchangeable definition of life, our new definition should have capability to evolve, but without losing its integrity. Furthermore, such a definition should be a living definition that will also alert us to its imperfections and knowledge gaps, so that we are constantly reminded that we must improve it. We hope to achieve this by applying systems thinking and analysis to defining life.

Life is a system of many parts which interact among themselves to create networks and other complex interactions. We need to identify those parts and their interactions that are both necessary and sufficient to adequately describe life. Furthermore, life is an extremely complex phenomenon, and is composed of not only parts but also of subsystems, which themselves have their own parts.

By treating life as a system, and by applying systems thinking/analysis approach to it, we propose this new definition of life, comprised of three parts. The first part defines life as it is on Earth. We modify and expand this definition to be astrobiologically significant by addressing the origin of life. Finally, we modify the definition to make it applicable to the putative ET life. Our definition thus comprises the three parts shown below.

4.3   OUR SYSTEMS-ORIENTED DEFINITION OF LIFE

Definition of life as it is on Earth:

> Life is a complex system comprising three subsystems – those of metabolism, information, and membranous compartment – and is a part of two larger systems, those of the environment, which supplies nutrients and energy, and biosphere, which, in conjunction with the environment, enables evolution.

By "complex" in this definition, we mean complexity by chemical diversity, organization, and established networks.

Please observe that we have defined "complexity" as applies to our definition. This takes out the ambiguity of what complexity is. We have not quantified this complexity. This may not be necessary since we have enough information about the complexity of life and its subsystems.

For any definition of life to be astrobiologically significant, it must include an explanation of the origin of life. We thus supplement the previous definition by adding the following sentence:

> Life originated in our Solar System by prebiotic chemical evolution, which led from heaps of prebiotic chemicals to prebiotic proto-life systems, which then underwent transition to life by an unknown process, for which, however, several hypotheses exist.

This part of definition is not as well developed as the previous part. The reason is that the chemical evolution that led to life and the transition from proto life to life are still at the hypothesis stage and have significant knowledge gaps.

To make the previous two parts applicable to extraterrestrial life, we modify them accordingly:

> Life is a complex system comprised of three subsystems – those of metabolism, information, and membranous compartment, **or their functional equivalents**, and is a part

of two larger systems, those of the environment, which supplies nutrients and energy, and biosphere, which, in conjunction with the environment, enables evolution.

We assume that extraterrestrial life is analogous to our life in a fundamental way. The key element of this definition is *"functionally equivalent"* as compared to the subsystems of life on Earth. This acknowledges a possibility that different evolutionary pathways on other planets may have selected different subsystems, but with functions that are same or equivalent as ours. The concept of "functional equivalency" may have been relevant also to the chemical evolution that led to the life on Earth, since it might have produced, as one example, alternative genetic systems that were later taken over by the present one (Yu et al., 2012). Still, the definition of ET life is the most hypothetical of all the parts of our definition, and most likely to change and improve as we hopefully discover such life.

However, our definition of life is not detailed enough to address all the intricacies of life. Only by studying the characteristics of life as a system, its subsystems, and interactions with the environment can we obtain a more complete picture. We show some examples in the subsequent chapters.

Life as a system has some characteristics in common with systems that are not alive, such as an engine. Life of an individual organism has a spatiotemporal character and a limited duration. As a system, it comprises parts that interact with each other in such a way to enable life to function. As time goes by, the parts become less and less functional and lose their capacity to contribute to the functioning of life's system, which also happens in an engine. Then, the system reaches a pivotal point, after which its resilience to the change drops, and the system ceases to exist. This would be analogous to an engine that eventually stops functioning. Life, however, has the option of reproduction which creates a way for the life to continue in time and space, via its offspring which comprises a new system with fully functional parts. In Section 4.4, we exploit this feature of life and consider a robotic system that would have such characteristics.

Life's subsystems are composed of subsidiary subsystems. For example, what we call "metabolism" includes two subsystems, catabolism and anabolism. Catabolism can produce ATP in a series of reactions in which food is broken down. ATP is the energy currency, which is then used in anabolism, to provide chemical energy for biosynthesis reactions to occur. The two subsystems, those of catabolism and anabolism, are thus linked via ATP (Bettelheim et al., 2010; Voet and Voet, 2011). This aspect of production and utilization of chemical energy is well known. Figure 4.1 shows schematically catabolism and anabolism. A more detailed look at the production of ATP is shown in the Krebs cycle, Figure 4.2.

The possibility exists that there was a primordial energy currency that was not ATP but some other high-energy precursor, such as pyrophosphate (Pasek et al., 2008; Pasek, 2020) or cyclophosphates (Britvin et al., 2021), which, however, functioned similarly.

Life is an out-of-equilibrium system, which must be maintained for life to function. If not, life reaches an equilibrium with the environment, which spells death. When life undergoes self-organization, its entropy goes down (more order). The relationship between the order and entropy is shown in Figure 4.3.

For the order in a system to be maintained, the environment becomes more disordered. This can be achieved only in a thermodynamically open system, in which both mass and heat are exchanged. This is presented in Figure 4.4.

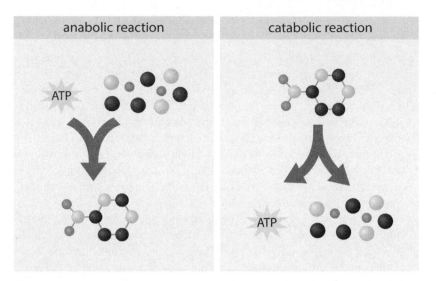

FIGURE 4.1 Metabolism involves both anabolic and catabolic reactions. (With permission from Shutterstock.)

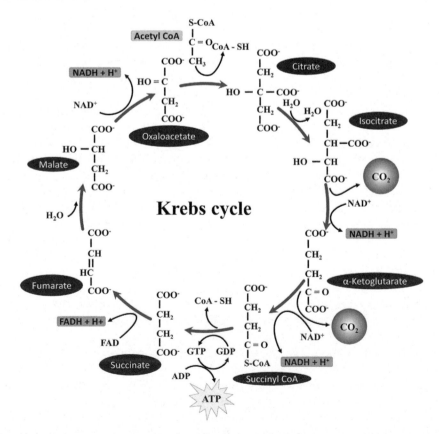

FIGURE 4.2 Krebs cycle. (With permission from Shutterstock.)

Energy, Entropy, the 2nd Law of Thermodynamics

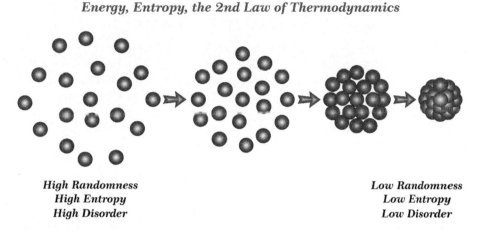

High Randomness
High Entropy
High Disorder

Low Randomness
Low Entropy
Low Disorder

FIGURE 4.3 A graphic presentation of energy, entropy, and the second law of thermodynamic. (With permission from Shutterstock.)

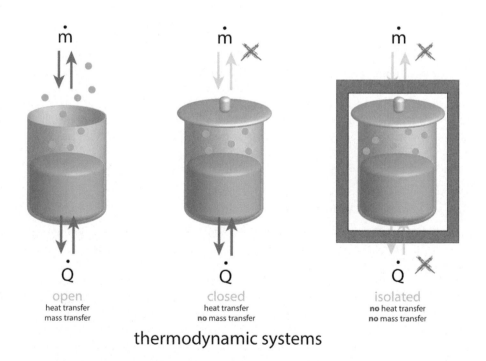

thermodynamic systems

FIGURE 4.4 Thermodynamically open, closed, and isolated systems. (With permission from Shutterstock.)

The processes by which protocells and primitive life established and sustained their out-of-equilibrium status are currently the subject of an intensive investigation.

4.4 LIFE AS A SYSTEM VS. ROBOTIC SYSTEMS: WHAT CAN WE LEARN FROM THE PARALLEL WITH THE MARTIAN ROVERS?

Life is a system that shares many of its characteristics with some systems that are not alive, such as the robotic systems, for example, the martian rovers. Just like the robotic systems, an individual organism has a limited duration. Both martian rovers and an organism comprise the parts that interact with each other in such a way to enable these systems to function. As time goes by, the parts become less and less functional and diminish their capacity to contribute to the functioning of these systems. Then, the systems reach a pivotal point, after which their resilience to the change drops, and they cease to exist. However, an organism as a system has the option of reproduction which creates a way for life to continue in time and space, via its offspring which comprises a new system with fully functional parts. The robotic systems in principle could also have such characteristics. Particularly important would be the design of mutations and the evolution of such systems. They would be needed for adaptation of robotic systems to the environment. In the case of the rovers, the conditions of the environment are characteristic of Mars, but the martian environment and habitability change over time. When we design a robotic system, whose purpose would be to travel in space to reach distant locations, we would need to build in a warning system that would prevent catastrophic collisions with space objects during space travel. Could we then build an eternal robotic life, or, more modestly, robotic life which would last long enough to reach another habitable planet outside the Solar System? Can we build in the information about our life on Earth for the robotic life to transmit to the habitable planets, and perhaps even start life like our own there? The discussion below is based on the following references: Lakdawalla (2018), NASA Mars2020 Rover Website, and Conley (2019). This coauthor (Clark) has personally worked with Mars lander and rover missions for several decades, and some of the following discussion is based on these experiences. This input does not include any classified information and does not require permission to publish.

4.4.1 System Description and Classification

A classic case of a robotic system is any complex entity with an engineering basis. One of the more challenging examples is that of a semi-autonomous rover sent to Mars to explore the planet for astrobiologically related goals, such as investigating whether the planet is or was habitable for life as we know it. Mars is today a cold and very dry desert environment, with periodic planet-encircling dust storms and no known surface reservoirs of liquid water. From an abundance of evidence, including dry riverbeds and lake beds, as well as numerous enrichments of water-soluble salts, it is clear that Mars at one time was warmer and decidedly wetter than its polar ices and widespread subsurface permafrost of today, and therefore habitable (e.g., Rivera-Valentín et al., 2021; Checinska Sielaff and Smith, 2019; Cockell et al., 2016).

Several NASA missions to Mars have landed rovers to roam the surface in search for detailed evidence of the history of habitability. The most recent rovers, "Curiosity" and "Perseverance," are nuclear powered and thereby not subject to degradation of solar cell power by the periodic dust storms. The Perseverance rover,

FIGURE 4.5   The Perseverance Mars rover with camera mast (a) and arm with instruments and drill (b) deployed vertically. (From Shutterstock, with permission.)

shown in Figure 4.5, may not appear streamlined. That is partly because the Mars rovers move very slowly and deliberately, slower than a person taking a stroll. The other reason is that even if there was a superhighway, a vehicle would not need to be sleek because the air on Mars is about one hundred times less dense than Earth's air, and hence there would not be significant air resistance even during high-speed travel.

Any rover is, however, a very complex system because of the many functions it must perform. Not only must such systems be mobile, but they must also be capable of operating sophisticated cameras and many other scientific instruments, including mapping spectrometers. They must also be capable of operating in a very dirty and dusty environment, with sharp rocks, cold temperatures, and strong UV exposure. To be most efficient, they must be able to drive themselves for portions of a traverse, including detection of positive hazards (rocks) or negative hazards (steep depressions or slopes) along the way. They must be able to communicate during intermittent over-flights by relay orbiters, and they must be able to detect and characterize problems that arise, then go into a "safe mode" state and transmit that information to Earth.

Modern rovers have such manifold capabilities that they manifest almost as a living organism. For example, a naïve putative martian who came upon such a machine might mistake it for a Marsonaut from Earth, once they observed it moving across the planet, and occasionally "talking" to the orbiters above. Rovers have stubby "legs" (with wheels for "feet"). They have an arm – what we call a robot arm; and some can pick up samples with a "hand" gripper or scoop. Some can "smell" with their mass spectrometers, gas chromatographs, or tunable laser spectrometers. Rovers take up portions of the soil or bore into rock and can then "eat" that by inserting it into their "mouth" (sample inlet to internal instruments). All can "see" with their 3D color-vision eyesight that extends into both the IR and UV. They can even vaporize small spots on nearby rocks with their laser beams. It is now even possible for such rovers to "hear" with an onboard microphone. And after they grind into a natural sample, they can "puff" away the powder to get a better look at the insides of some intriguing rock or sediment. Their antennas allow them to "talk" to orbiters as they scream by overhead, or even talk directly to Earth, albeit at a much slower rate.

What is the system that enables all these capabilities? Fundamentally, it is quite straightforward. In implementation, it is quite complex. The key to designing the next rover to go to Mars is to first define and analyze its functional requirements as a system. Then, using standard practices in system design, analysis, and control, the engineers can create the rover that will survive and perform the desired tasks.

The key to engineering success is to employ the techniques of systems engineering to set requirements for performance, divide the overall system into a set of subsystems, each with a well-specified functionality, and set how these subsystems must interface and operate within the overall system.

Conventionally, experienced engineers will divide up the system into already well-known subsystems, for which they often have a long heritage of experience. Within each major subsystem can be many smaller devices or subsystems. For example, a subsystem of "mechanisms" can include motors for moving the wheels, an arm, an antenna, etc. These motors must survive in the harsh martian environment, with its gritty dust and extremely cold temperatures at nighttime. Each motor needs power circuits to drive it, and controls for speed. Heaters may be added to these motors as well as to many other subsystems to make sure none of the components ever becomes so cold that it cannot operate or even survive.

Two early rovers, known as "Spirit" and "Opportunity," eventually succumbed after years of operation once their solar cells became so dusty that they no longer could power the electrical heaters to avoid "freezing" damage to electronic circuitry due to thermal stresses. Nuclear-powered rovers do not have this disadvantage, although the thermoelectric devices that convert the heat energy from the radioisotope source slowly degrade and eventually will succumb to the ravages of time. Like living organisms, such rovers have a "birth" (when they emerge from their entry capsule and land on Mars) and a "death" (when their computers or communications subsystems no longer have enough power to operate).

These considerations lead to the designation of additional subsystems, such as the Power subsystem. Then there is the one for "Control and Data," i.e., the central computer subsystem with its data storage, command sequencing, and analysis capabilities. And there is a communications subsystem, for the antennas, radio-frequency generators, and modulation equipment.

With this in mind, let us first delve deeper into what are all the subsystems that engineers must deal with and develop a framework that may be applicable more generally to a wide variety of systems.

Many of these functions involve a transportation or re-location of things. Transporting matter is conveyance, or in our examples, the subsystems of "mobility" and "input/output." Transportation of information involves "communications." Supply and transport of energy is called "power." And transporting instructions for control are specified as "commands."

In Figure 4.6, we provide a typical subsystem block diagram for an engineering system such as our planetary rover. This incorporates several of the subsystems

SYSTEM FUNCTIONAL BLOCK DIAGRAM

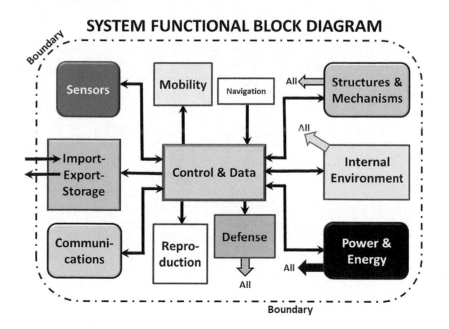

FIGURE 4.6 Functions performed by the various subsystems within a complex engineering system, such as a Mars rover (or automobile, airplane, chemical processing plant, rocket, spacecraft, biological organisms, etc.). (Produced by coauthor Benton Clark, no permission needed.)

to which we have already alluded. Altogether there are 13 identified functions (including the Interconnects, signified by the network of lines and arrows, and the Boundary).

In general, the overall function of the system is allowed to constrain supporting functions of the subsystems of the subsystems. However, there are often overriding top-level requirements that dictate starting points. For example, our Mars rover must operate from a single power source, a specific "space-rated" nuclear battery that is only available in one size. And the rover must physically fit inside its entry aeroshell which, in turn, with its cruise stage must be compatible with the inside dimensions of the heat shield of an existing, affordable launch vehicle (rocket). These types of requirements immediately impose maximum size limits on the rover and the maximum amount of energy it will have to accomplish its daily tasks. Furthermore, it must be able to survive in the cold, ultra-dry martian environment and its rough terrain. With these factors in mind, let us examine in more detail the various subsystem functions that must be performed to create a mission that satisfies the science objectives within a prescribed cost/benefit trade study. For the Perseverance rover, the following functions can be further examined in terms of their capabilities and limitations. Except in cases discussed below, where a specific reference is cited, details relate to the Curiosity rover and its reference (Lakdawalla, 2018), from which the Perseverance rover is derived, or to the appropriate NASA website (https://mars. nasa.gov/mars2020/spacecraft/rover/).

4.4.2 POWER AND ENERGY

The rover's nuclear battery is fueled with an artificial radioactive element, plutonium-238, created in a nuclear reactor by reactions with neutrons. The hundred-pound MMRTG (Multi-Mission Radioisotope Thermoelectric Generator) converts the 2,000 W of thermal energy its plutonium generates using a thermoelectric converter unit to create about 110 watts of electrical power. However, to run all the equipment and thermal control system on Perseverance can require more power for many consecutive hours when certain instruments are powered on. The way this is enabled is to charge two lithium-ion batteries, having ~1 kWh storage capacity each, to provide higher power levels when needed. Recharging the batteries can occur when other activities are resulting in less power consumption, especially through the use of modes variously called "sleep," "nap," and "deep-sleep" when the rover has drastically reduced how much equipment it has powered on. Although the half-life of the Pu-238 radioactive material is almost 88 years, the "operational life" of the rover is expected to be only 14–20 years, depending on the degradation rate of the thermoelectric converters, and the ability of a cost-acceptable mission to collect science data with much lower operating power.

4.4.3 CONTROL AND DATA

This is a block function implemented by a space-rated computer using special, streamlined software that seldom if ever "crashes" and is isolated from interplay with other computer systems (e.g., the Internet) so that "hacking" is not possible. The radiation-hardened RAD 750 PowerPC computer can operate at up to a 200 MHz rate, and includes 2 GBy of flash memory. It utilizes a special operating system, VxWorks, which is designed to operate for years without necessarily being re-booted.

"Commands" is the set of controlling directions that specify which subsystems of the rover are active, and schedule what and when they are to accomplish each task. Science instruments are, for example, generally turned off until it is an opportune time to gather data (e.g., daytime for cameras, at favorable sun angles).

Onboard memory can store a few days of rover-generated data, so that once a specific data set from the instruments and engineering housekeeping streams is received and verified on Earth, it can be marked for deletion to make room for subsequent data acquisition. To conserve energy, the Control function is updated daily with a new script for the next one to three sols so that instruments are only powered on when the taking of new data has been approved and scheduled.

4.4.4 SENSORS

Sensors include many functions, and for a scientific mission, some need to ingest samples in order to analyze them appropriately. For example, samples can be processed chemically, with reagents, or physically, by heating to release volatile constituents. Stereo cameras can be passive imagers, but also with bandpass filters to separate out portions of the reflected wavelength spectrum to discriminate among minerals. On Perseverance, a mapping Raman spectrometer shines laser light onto samples to stimulate diagnostic emissions by minerals, salts, and organic compounds. The SuperCam instrument employs several sub-instruments to look for Raman and fluorescence effects from its laser stimulation as well as passive spectrometry of sunlight reflectance. Its laser can also be used to vaporize tiny volumes of rock to determine elemental constituents, while a microphone records acoustic signals from the impulse. The PIXL instrument (Planetary Instrument for X-ray Lithochemistry) uses a focused beam of low-energy X-rays to stimulate X-ray fluorescence for creating maps in 0.12 mm steps for the elements in various minerals within rock and soil targets. In addition, an array of meteorological sensors measures changes in temperature, pressure, humidity, UV, dust, and winds in the atmosphere. A ground-penetrating radar unit detects changes in subsurface layering.

Altogether, nearly two dozen cameras were flown to Mars with the Perseverance mission. In addition to seven different science cameras, there are the engineering cameras, including six hazard monitoring cameras (to detect rocks, holes, and soft soil) and two-color stereo navigation cameras scouting for landmarks for navigation purposes. Other cameras monitored the landing, and a sample camera images the end of each core tube after a drill operation.

About 200 thermal sensors spread out over the rover monitor temperatures at various locations so that any overheating or problematic lower temperatures can be detected, and appropriate measures be taken if needed.

4.4.5 STRUCTURES AND MECHANISMS

The rover body is primarily constructed of aluminum alloy and includes the WEB (Warm Electronics Box). Science instruments are provided by the Principal Investigators, who have responsibility of following the requirements for robustness, choices of materials, and providing their own radiation shielding if their instruments require it. The ruggedness of each engineering and science "black box" on the rover is tested separately for survival of the vibration and shock levels during launch and landing.

4.4.6 IMPORT-EXPORT-STORAGE

It might not seem likely that a planetary rover would "export" any mass, but there are several examples of just that. For example, a deployment system was used to set the helicopter down on the Mars surface, and some items were jettisoned as part of that maneuver. Of course, data is transmitted, and the information eventually erased, but that is covered under "Communications" and does not result in changes in mass.

Although the system boundary is meant to tie together the essential functions, it generally must be porous to transport of needed products inside and transport waste products outside. For a Mars rover, the import function is the obtaining of rock and soil samples, so that selected ones can be stored for later transportation to Earth (by a different system).

There generally are no mass waste products from a rover because it does not use chemical fuels or other consumables.

However, to send a rover to Mars involves very massive and complex systems. The normal sequence is launch into Earth orbit by a huge rocket at the precise time → departure from Earth by firing another rocket at a precise location in Earth orbit → coast to Mars, with occasional trajectory "tweaks" using very small rocket-engines ("thrusters") → encounter with the martian atmosphere at an extremely high speed, greater than 12,000 miles/hour. The next sequence is termed EDL (Entry-Descent-Landing). The process can be quite complicated because just after the protective capsule is slowed sufficiently, a parachute is deployed while still going above the speed of sound, with the parachute slowing the vehicle further but also enabling the heat shield to be dropped off and thus freeing the payload to fire retro-thrusters to slow further before touching down. In the case of the Perseverance rover, this last stage was especially complex, employing a "sky crane" that lowered the rover on cables to keep the engines high above the ground and then flying away after the rover was placed on the ground. This avoided the large amount of dust and rocks that would be kicked up by the engines if they were too close to the soil.

Huge amounts of propellants were therefore expended by the original launching rocket, as well as by the cruise stage during trajectory corrections and during descent of the rover and its "sky crane" that gently lowered it to the martian surface before flying away and crashing elsewhere onto the martian surface. In addition to the sky crane, other items were discarded during EDL, including the heat shield and back aeroshells, and the parachute.

Finally, Perseverance is unique in its ability to obtain core tube samples of hard rocks, sediments, and soils, and then capable of deploying them to the surface for later acquisition for transfer into the rocket which can return them to Earth.

The acquisition of the core samples is an example of Import, and also of Storage since the filled tubes are stored until time to transfer them. The tubes themselves were thoroughly cleaned and the rover itself was also cleaned of all potential contaminants, especially diligent for removing organic compounds which could contaminate the samples and give false indications of what materials are actually on Mars.

A final requirement significantly drove the efforts and costs of producing an acceptable rover mission. That was to clean any organisms, especially microbial spores, from the rover and other hardware that landed on Mars. This was to satisfy requirements of Planetary Protection, i.e., to avoid "forward contamination" of Mars with viable organisms from Earth (Conley, 2019). Not only is this important from the standpoint of maintaining the samples pristine, but also for preventing the possibility of some extremely hardy microbe from reproducing on the martian surface and then being spread by wind to potentially contaminate the entire planet and forever compromise any future discoveries of the presence of organisms or their remains indicating there is now or once was once life on Mars.

4.4.7   Communications

The rover does not communicate continuously with Earth. Ironically, the martian day (its "sol") is only 3% longer than our day on Earth. There is no fundamental reason for this close correspondence, and most other objects in the Solar System rotate at much different rates than we and Mars do, ranging from less than 1 hour (for asteroids) to many months (for tidally locked objects, like Venus, Mercury, and moons of the giant planets). However, that extra 39.5 minutes of the martian day wreaks havoc when we Earthlings attempt to match the martian day in order to synchronize operations. After a couple of weeks, the morning where the rover is (e.g., Jezero crater) turns out to be midnight where the rover operations team is (California). All the people involved in the mission are susceptible to jet-lag phenomena or changing-shift syndrome, and eventually no longer can perform at peak efficiency. Thus, the missions only attempt to operate on Mars-time for the first few weeks or months, and then switch to a methodology of pre-planning a few sols ahead as much as possible on Earth time.

Communication with the rover is not continuous anyway, because in order to transmit the large amounts of science data that are being collected each sol, it is much more efficient to communicate when one of the four available orbiters around Mars passes within UHF transmitter range of the rover. Since these orbits are known with extreme position, it is possible to predict days or weeks ahead at which time each orbiter will become available for a few minutes of receiving transmissions from the rover. The selected orbiter then re-transmits the data to Earth using its large dish-type antenna and its higher energy from its large solar arrays to telecommunicate over S-band frequencies to pre-pointed, large-dish tracking antennas on Earth. This process is termed "downlink," and the sending of the command sequence for upcoming sols of operation up to the rover is termed the "uplink" process. The one key uplink process that goes directly to Earth from the rover every morning is a short sequence known as the "beep," which signals that the rover has fully received its command load for that sol and is operating normally, and therefore will proceed with the plan. Since there are three NASA orbiters plus one ESA (European Space Agency) orbiter that can support relays, at least one communication link occurs each sol, and often two may occur.

4.4.8   Internal Environment

This subsystem includes the thermal control function, which is accomplished in engineering systems not only by active heating (or cooling, for spectral detectors) but also by various "passive" means, such as coatings that emphasize absorptivity, emissivity, or reflectivity depending on whether it is desired to import solar thermal energy or to reject it. Using louvered shades, it is possible to turn such coatings into active devices, responding to changes in thermal conditions. Heat energy can also be transported from one area to another (e.g., from a unit with high power dissipation to a radiator) with solid conductors (heavy!) or heat pipes which take advantage of closed-loop liquid-vapor transformations.

The nuclear-battery-powered rovers have available large amounts of waste heat from the MMRTG, which can be used effectively to warm the electronic equipment that otherwise would be subjected to the cold martian atmosphere. Mars is quite cold, but the Perseverance critical electronics are enclosed in the thermally insulated WEB. Using Freon-11 (banned on Earth to protect the ozone layer and its shielding of dangerous UV sunlight), the coolant is pumped between warm plates heated by the waste heat from the MMRTG to the electronics equipment ("avionics") in the WEB. If the avionics area is too warm, the Freon coolant can be pumped from that area to "cold plates" exposed to the cold martian atmosphere. This thermal control system operates independently of the central computer control system because it is critical for survival of various electronic units, including the computers themselves.

The reason this is important is that although electronic equipment is far more tolerant of temperature changes than biological systems, there are extremes beyond which performance cannot stay within needed limits. And at even further temperature excursions, either too high or too low, most systems can be irreversibly damaged. Thus, spacecraft and our rover are tested in thermal chambers to verify their proper operation at expected temperature extremes. Other items need to be heated electrically, such as the motors that drive the wheels. This is all part of the generic subsystem of "Internal Environment," which controls all critical temperatures.

The interior of the rover and WEB are not pressurized, however. Thus, all equipment must be compatible with the very low pressure on Mars which enables electrical discharges for higher circuit voltages. In addition, the electronics in each box must be tested to assure its accuracy in the stressful operations on Mars.

Once landed and after checkout tests, the Perseverance rover was free to roam on its own. In the block diagram, this self-propelling is termed "Mobility" and the subsystem that helps guide it and record its whereabouts is "Navigation." Following more checkout, the rover carefully deployed the small helicopter it had brought along. The "Ingenuity" helicopter is its own system but does use the rover to radio the data it collects, including images of the terrain it flies over, to the rover for later transmission to Earth.

4.4.9 MOBILITY

The ability to relocate finite objects is one hallmark of systems. Matter may need to be conveyed from one location to another. The rover's six wheels are independently powered and steered, and not freewheeling. In fact, when motors are off, brakes are applied to the wheels, and if a motor fails, that wheel must be dragged by the force of the other wheels. The sophisticated "rocker-bogie" suspension system allows pivoting in ways designed to keep all six wheels on the ground even when driving over rocks and very uneven ground.

4.4.10 NAVIGATION

Fore-aft tilt and sideways tilts are readily measured relative to the gravity vector by the components in the rover's IMU (Inertial Measurement Unit). Because the IMU includes a gyroscopic compass, it can get a reading of which direction the rover is pointing along each drive segment. However, this reading becomes less and less accurate over time after the IMU is calibrated, but by checking the position of the sun taken at a selected time of day, using a camera image, the rover drivers can pin down its actual compass heading at the time of the picture. This can then be used to periodically recalibrate the gyrocompass.

It is also important to keep track of where the rover has driven in order to understand which direction a camera was pointing when it took each picture, and to determine what is the geologic setting relative to all the images being taken from the ground as well as the images that have been taken from orbit. With stereo vision and documenting the distances driven by counting wheel rotations, a good idea of where the rover is actually located at any given time can be reconstructed. This can be correlated with the down-looking images taken by the orbiters to identify various obvious local features (e.g., a peak or cliff) to determine where the rover is relative to those comprehensive pictures that make up the map of Mars.

4.4.11 REPRODUCTION

Without the design engineers, the rover would never have been created. The engineers can even reproduce the rover, as desired (e.g., there were two identical MER rovers) or enabled by the financial sponsors (taxpayers, via NASA). Indeed, the Perseverance rover is a "copy" of the Curiosity rover, to save costs. However, it was prudent to upgrade the wheels to make them more robust, and nearly all the science instruments are different in size, shape, mass, and operating requirements. Furthermore, Perseverance has the additional requirement to use a coring drill to obtain samples that are stored in each of the core tubes used. These changes were significant, but nonetheless many engineering subsystems and the overall rover design are primarily a faithful reproduction of the Curiosity design.

In previous decades, before the computer revolution, designs of space hardware existed in hardcopy of hand-drawn designs of not only the dimensions of components to be built but also an engineer's instructions of how to fabricate and assemble them. These "blueprints" or "drawings" are now mostly obsolete. Most designs are accomplished with the aid of computer graphics software, and those designs can be ported over to numerical-control machines which can produce the desired parts and assemblies. In older times, when some item was being fabricated, the drawings on paper were followed, but if a change was needed quickly, the drawings were "red-lined." It was still possible to manufacture a subsequent copy from the red-lined drawings, if budgets were insufficient to re-issue the paperwork with corrections. It is interesting to think of this process as analogous to biological evolution, where the original drawings are the "DNA" of the machine, and the red lines are the "mutations" of the design.

4.4.12 DEFENSE

Although the rover would presumably never be under explicit attack, there are nonetheless several safeguards that are important to defend against adverse natural or antagonistic actions. A rover therefore needs to know how much it is tilted because of the danger of tipover, or slippage on a steep slope. Both fore-aft tilt and sideways tilts are needed, and these are readily measured relative to the gravity vector by the rover's IMU.

For example, an enemy of moving parts is any contaminant that could cause jam-ups. Therefore, there are seals to prevent dust from reaching sensitive mechanisms with close tolerances of moving surfaces. Protecting against the ultra-fine-grained martian dust is one concern. Another is in the event of FOD (Foreign Object Debris) being carried onboard and subsequently causing interference with a valve or other mechanical device. There have been cases of an entire rocket failing because a small contaminant object got in the wrong place. The FOD can occur when, during the manufacturing process, some small item or residue becomes hidden in a crevice or corner of the rover, only to reappear when the system is in zero gravity on the way to Mars, or when the shock of entry or landing causes it to reemerge.

An important property tested for each black box is whether it emits radio waves, or conducts high-frequency electrical noise on its cable connections. Each unit must pass tests to assure there is no significant level of EMI (Electromagnetic Interference) generated. Likewise, all units must demonstrate that they are also not unduly sensitive to such interferences, and hence have EMC (Electromagnetic Compatibility) resilience.

Space radiation, in the form of galactic cosmic rays and also high-energy particles occasionally emitted during certain types of solar storms, can also upset or damage electronic circuits. The remedy is to use only electronic piece parts that are designed to resist the damage by ionizing radiation, and/or to provide shadow shielding or thick walls of their enclosures to shield out the majority of the damaging radiation.

Some instruments cannot safely be exposed to the direct rays of the sun (as with the human eye, or outer part of the human body which is susceptible to skin cancer produced by excess exposure to UV rays from the sun). The usual remedy is to avoid pointing such instruments in the direction of the current location of the sun, which depends on accurate knowledge of the orientation of the rover itself, as well as taking into account the time of day.

The MMRTG is heavy because its eight modules containing the hazardous radio-active plutonium are each encased in carbon-fiber aeroshells which, in the event of a severe launch accident, would survive the heating of reentry and the impact onto the ground, keeping the dangerous material fully contained.

Software updates and command scripts must be protected against hackers by extensive protocols on the ground. Also, the flight computer only accepts changes that have the proper authentication codes as well as a checksum and other methods to assure that each new S/W load is exactly as sent from Earth.

4.4.13 INTERCONNECTS

In the overall block diagram, there are many lines and arrows, showing the inter-connections between the major blocks. Within each block, of course, there are also smaller networks that interconnect its own sub-components. These network paths are especially used to distribute energy and information. In the rover, this is by use of electrical wiring, in one or more "wire harnesses" or "cable assemblies." These cables are often surprisingly complex and heavy. Although data could be transferred wirelessly between the functional blocks, in rovers and most critical engineering systems to date this is generally accomplished by hard wiring. Furthermore, for critical signals, there is often a metallic sheath covering over the insulating coating of the data wires to prevent inadvertent signals from becoming induced onto the data line.

Within something as small as a typical prokaryotic cell, interconnecting networks are not necessary because the diffusion time for molecular transport is so very short due to the small submicron distances involved. Thus, energy, mass, and information can all be transferred quickly. For more complex cells, such as eukaryotes, there are transfers that are facilitated by structural pathways, such as protons transported across the mitochondrial membranes. And for multicellular systems, including the human one, there can be elaborate network systems such as the neural network of neurons connected by synapses, the cardiovascular network of blood vessels, and the subsidiary network of lymph channels, which eventually feed into the bloodstream.

4.4.14 BOUNDARY

A system has a boundary. The boundary may include a membrane, wall, or fence, but in general is it not impervious to everything.

In the macroscopic world, a fence is a boundary, but the fence may be porous. For example, a wire fence may prevent the escape of large objects, such as cows, sheep, or cars. Yet, smaller entities such as a rabbit or mouse may be able to pass through the fence. And an agile object, such as a deer, may leap over the fence with ease. And birds and butterflies may fly over or even fly through the fence. Yet, the fence is performing its primary purpose.

In the microscopic world, the boundary of a cell is a membrane, but these are not like plastic sacks with no holes or openings. This membrane serves to keep inside the ribosomes, the DNA, and most proteins and other macromolecules so that they can interact with each other, while at the same time allowing passage for feedstock and waste molecules, as well as active transport of certain ions in solution.

Some systems, such as an automobile, are not encased in a cocoon. Instead, it has an outer limit, with a complexity of convex and concave areas. The "membrane" around this outer surface is made up of paints, plastics, and metal plating (e.g., chrome). These coatings protect the underlying metals from corrosion. Glass material may have a coating also, including plastic to prevent shattering (windshield).

The boundary on a spacecraft is generally established by hard structure, whether it is a body encasing certain equipment, or simply the exteriors of propellant tanks, rocket engines, solar arrays, or antennas. However, there is also often a membranous layer, composed of several layers of aluminized Kapton or other plastic film. These so-called MLI (Multi-Layer Insulation) blankets are used to isolate the interiors from the heat flux from the sun which, un-attenuated by the atmosphere, can produce large unwanted rises in temperature.

4.4.15 Lessons Learned from Systems Engineering

"Requirements" are what designers and engineers impose on a system when it is a top-down design. Biological organisms are based on designs that evolved from the bottom-up, and hence are not necessarily optimized for all functions the organism performs.

For example, the hawk may have an eagle-eye ability to spot small rodents from high altitudes, but its legs and feet have marginal capabilities for walking yet can have powerful, razer-sharp claws. Indeed, many birds just hop (although seagulls can "run" and the ostrich and emu can run faster than the fastest human can sprint). And, of course, most birds fly much faster than we could ever run. They also react much faster to challenges, and an entire flock of birds or fish can change direction totally together while we are only ruminating and contemplating what to do.

4.4.16 Optimization

It is important to keep in mind, however, that it is not necessary to optimize each individual component or subsystem in order for the system to function adequately. Or, for that matter, for the system to function very well in spite of infirmities or some non-optimized subsystems. The most essential requirement is that the subsystem be able to function sufficiently well that the overall system can perform for survival and mission accomplishment. "Better is the enemy of good-enough" is a mantra of engineers when developing a new system.

At the bottom line, organisms only need to perform optimally with respect to *survival* in the environment in which they exist. There is no need for the ostrich, dumb compared to domestic dogs and cats, to be more intelligent when it can run far to find food, fast to escape predators, and is equipped with sharp claws on its powerful legs that they can even kill a lion if necessary.

4.4.17 Operations

A system's performance generally depends upon its Operators. Perhaps surprisingly, since the public media often states otherwise, it is the engineers, not the scientists, who actually "drive" and otherwise operate the Mars rovers. It is the creation of engineers, and they are responsible for its health and safety, especially considering the financial investment and diversion of technical resources needed to develop and operate it.

The selected scientists are, however, also part of this overall System formed for purposes of conducting on-the-ground exploration of Mars. And it is a group of impartial scientist peers who originally selected which instruments would be accommodated on the rover.

Once on Mars, operating the rover by a combination of consensus, leadership, and pre-approved goals and objectives, the science team proposes the locations to explore and which scientific instruments to operate, while the engineers concentrate on safety and mission success. In the unique case of the Perseverance rover, the science team also decides which samples to take as candidates for future return to Earth.

4.4.18 Distributed vs Integrated Subsystems

In reality, a subsystem can be "distributed," or it could be an "integrated" subsystem. An example is the use of small computers (or CPUs) inside instruments that talk to the central computer. These smaller computers typically orchestrate the detailed operation of the instrument. Another example is thermal control, which is typically spread out over the entire mechanical structure in order to guarantee that temperatures do not go outside the upper and lower bounds and that each box has been verified to be safe for its operation.

4.4.19 Testing

Often, human engineers will design subsystems to be quite independent because it facilitates individual testing (and furthermore eliminates joint levels of responsibility). This way, the key subsystems can be developed in parallel, by different dedicated engineering teams. As long as the lowest-level requirements are specified in sufficient detail, the overall system should come together and function properly. This is how you put together a detailed design for a rover, or a spacecraft, or a rocket, or for that matter, an automobile, and then expect it to perform properly. This is the epitome of the top-down approach for development of a new system.

Typically, the requirements hierarchy for a new system can span 4, 5, or even 6 sub-levels, with ever-greater specificity until there are literally several hundred or a thousand specific requirements, each of which must be verified or validated by test or worst-case analysis. To make sure all the requirements are being met, the engineers and technicians conduct tests at various points of integration. It starts at the piece-part level, for items such as transistors, capacitors, relays, and integrated circuits. Then, at the circuit-board level, culminating in performance tests when all the needed circuits are assembled into a functional unit. At this point, a test unit is built and subjected to various environmental tests to assure it will operate properly in the environment in which it is ultimately expected to perform. And even though these tests provide assurance at the component and "box" level, the final assembled rover is ultimately put through a battery of environmental tests to verify its survivability and operability. This includes high intensity vibration and shock tests ("shake, rattle, and roll testing") to simulate the transient mechanical forces during launch and high-speed entry into the martian atmosphere. It also includes tests somewhat beyond the highest temperatures and lowest temperatures expected to be experienced in the cold, dry, low-pressure surface of Mars. Previous tests will also have exposed similar components to the level of radiation expected for a 15-year or longer exposure in space and on Mars.

With biological systems, things are different. There are no detailed specifications waiting to be fulfilled. There are no isolated subsystem design, development, and test processes. Rather, all its subsystems must develop more or less in lockstep, both in the short term of reproduction and in the longer term of evolution. As a consequence, there is much less isolation of functionality. The subsystems are often more complicated and intertwined. Analysis by system modeling becomes far more challenging.

Also in biological organisms, multi-functionality of various parts and subsystems is often common, although it leads to complexities that expose the overall system to detrimental responses to various challenges. For example, the tongue. In animals, it not only helps move food around in the mouth to position it for grinding by the back teeth, but it also initiates swallowing and is sometimes the primary method of capturing and transporting food into the mouth. Lined with identification sensors (tastebuds), it provides the sensory signaling pathway from food to brain that gives the approval for keeping rather than disgorging the "food." For we humans, the tongue is also indispensable for efficient, meaningful communication with our compadres. Speech without an active tongue is extremely difficult, and singing is more or less a non-starter. The tongue is intimately involved in the act of swallowing. Even without vocal cords, the tongue can convey messages, such as hunger (licking of lips) or disgust (sticking-out-the-tongue). Thus, in our line-up of subsystems, the tongue is involved in one degree or another in Import, Communication, Sensors, and Energy distribution. And for these many reasons, life without a well-functioning tongue can be extremely daunting.

A number of precautionary actions during the development or actual operation of a new system may be taken as part of a disciplined approach to understanding the characteristics, as well as strengths and weaknesses of that system. If a system, subsystem, component, or even a single piece part fails to preform as intended during development and testing, that item must be subjected to extensive "failure analysis" to determine the "root cause" of the failure. Once found, a decision is made whether to correct or counter-act that cause of the problem, or whether to replace that item altogether with something functionally equivalent. Such an approach is difficult, given that the assembly and test of the whole system is on a timeline that has been previously planned in great detail. Prudent planning actually places strategically selected increments of time as "reserve" during which catch-up to the planned schedule is possible. This is not always practical, because some failures, faults, or late deliveries cannot be adequately predicted or allowed for. Unforeseen circumstances are one reason that the development cycle for new spacecraft, including rovers, can be anywhere from 3 to 6 years, depending upon the envisioned complexity and ambitiousness. Often, for conservatism, it is desired to implement a high degree of "heritage," i.e., re-use of simpler but proven subsystems and/or their components. Such conservatism obfuscates acceleration of technology, but in the high-stakes costs of such systems, it is often a judicious strategy. "Keep it Simple, Stupid," (KISS) is another mantra often heard in design circles. This does conflict with the natural desire of each creative engineer to improve or even depart significantly from previous versions of their subsystem. Innovation can be challenging, invigorating, and refreshing, but potentially risky to the whole project.

Biological life is held to a rigid employment of previous designs, because only small changes are likely to lead to a viable organism. However, the numbers of possible biological iterations are enormous compared to the limited re-inventions of engineered systems, given that life has had millions of versions of iterated organisms of any given species' design.

To fully understand the system that a particular organism represents, it is necessary to understand each subsystem and its relationships to the other subsystems that make it up. Through the use of genetic engineering, it is now possible to artificially alter in minor or major ways the underlying design of one or more subsystems that make up a given organism.

4.4.20 OPERATING MODES

Most systems will have more than one operating mode, each with some emphasis on a particular near-term objective by choice of which subsystems are turned on or being used more intensively. For pursuing science objectives, modes such as "Proactive," "Reactive," and "Discovery" are appropriate. For maintenance of engineering health and safety, modes such as "Regenerative," "Repair," and "Survival" are favored. Some example operating Modes are shown in Table 4.1.

TABLE 4.1
Modes of Activities

Mode	Activities	Rover Examples	Microbiological Examples
Proactive	Pre-planned activity based on defined goals	Drive to target; deploy arm to analyze or process sample (abrade or drill)	Flagella or pili motility
Reactive	Respond to expected or plausible events	Detect and drive around rock hazard; halt if tilt exceeded	Chemo- or phototaxis responses
Discovery	Exploratory activities without explicit expectations	Spirit rover excavation of Fe-sulfate soil; opportunity rover turnover of Mn-rich rocks	Random-walk motility
Contemplative	Analysis of data and comparison with hypotheses	Science team on Earth analyzes results and formulates new approaches and targets	Quorum sensing
Regenerative	Reduced activity to return to a more stable state	Low-power operational modes to allow recharging of battery: "nap; sleep; deep-sleep"	Stasis under adverse conditions
Repair or replace	Partial or total correction of a malfunctioning item	Alter software in an instrument; switchover to a backup computer or telecomm unit	DNA repair; highly redundant ribosomes and RNAs
Survival (SAFE)	Depart from current mode, into a survival mode	Health status of subsystems; power-off nonessential units; avoid excessive tilt, sharp rocks, quicksand; favorable solar orientation	Toxin chemotaxis; spore formation; enter cryptobiosis
Reproductive	Creation of a near-copy of the present system	Re-use of "heritage" designs (Perseverance rover is a "copy" of the Curiosity rover)	Growth via replication and binary fission

The survival mode is more generally referred to as "SAFE Mode," and involves responding to some detected anomaly by halting the current operations as well as turning off all nonessential units and then transmitting engineering housekeeping and other diagnostic data to the ground. Once the engineers diagnose the state of the rover to determine the criticality of the "fault" status, they send via the uplink commands an alternative sequence of activities that diagnoses the problem. Often, the issue is a minor one that can be readily remedied. Other times, some operation of some particular device may need to be avoided until the problem can be understood and rectified in the future.

One internal communication function is that for each of the subsystems, where appropriate, to pass along data that provides information on what is their status and their "health," i.e., is that subsystem performing as intended? Whenever a payload instrument or other subsystem is experiencing some anomaly in expected performance, that information is communicated to the Control subsystem.

Quite typically, unless the errant unit is vital to survival of the overall system, that unit will be powered off and "marked sick" until analysis by human interventionists can ascertain what is the problem, and how to fix it (or ignore it, if expedient).

Similarly, when a human being is not at optimum performance because of illness or some other impediment, it is typically the case that strenuous functions are postponed until the anomaly is resolved. Bedrest or other lowering of functional activity is often part of the solution. Lethargy is, of course, also one of the natural responses of the human system to certain adversities, such as sickness with a virus.

4.4.21   Perturbations and Stability

A well-functioning system can operate effectively and efficiently not only in its nominal modes but also can recover fully from perturbations. However, every system is vulnerable to perturbations, which are outside certain limits and which may cause minor, major, or even catastrophic effects on its functionality.

There are two general end-members in the classes of perturbations: fully reversible versus totally non-reversible. In the first case, the system retains all its previous functionality. In the second case, it permanently loses some portion of functionality. The real world is, however, seldom so black and white.

Rather, there are also perturbations that induce hysteresis, which is the effect that the system does not readily, if ever, return to its former state once the perturbation is over and the environment reverts to its previous condition. An example is when a liquid is super-cooled and remains fluid, but then suddenly crystalizes into its solid form. Once in the solid phase, if the temperature is raised, the solid does not begin to return to the liquid state (i.e., melt) until an even higher temperature is imposed. Another example would be when a unit is overheated to be permanently damaged such that it does not fully recover when the temperature is lowered.

A large fraction of chemical reactions exhibit this irreversibility, because they may occur sluggishly due to the barrier of "activation energy," but once two molecules have reacted to produce a third molecule, that third molecule may be stable at lower temperatures and refuse to "disproportionate" back into the two original reactants.

When hysteresis is so severe that the reversion to the original state can never be achieved, we say that the system has passed its "tipping point." The obvious example from physics is when an object is tilted beyond the angle at which its center of mass is past the edge upon which is resting, and tips over. Once it falls, how does it get up?

This is one of the central dilemmas with robotics – if it is overturned, can the robot right itself? Thus, our Mars rover must always avoid driving on slopes that exceed its tilt limit because if it did in fact turn over, it would have no means to aright itself. True, such rovers have arms. But usually only one strong arm, and typically not strong enough to lift the weight of the rover. We humans can contort ourselves, and use a combination of arms and legs to get back to a position from which we can stand up again, especially if there is a nearby object such as a chair or door handle to assist with the maneuver. But most robotic machines are helpless if you turn them upside down, and those that are designed to recover from turnover are generally not nearly as efficient at their central tasks and purpose. For this reason, most industrial fabrication robots are single arms or combinations of arms, each of which is strongly anchored to the factory floor. And most transportation robots, such as in an automated warehouse, have a low center of gravity relative to the wheels on which they rove, as well as smooth, flat floors.

Our Mars rovers actually have six wheels, and the "Spirit" Mars rover actually performed for several years after one of its wheels became inoperable, by simply dragging along that wheel with the driving traction that the other five wheels were able to produce.

4.4.22   SYSTEM CHARACTERIZATION AND KNOWLEDGE

It is typically hopelessly impractical to know every little detail about a system and how it will react to the variety of environments it may encounter. For example, although the advanced Mars rover missions are very well supported, with hundreds of engineers having been involved, there are numerous times when a rover will not accomplish the daily plan programmed into it. Although some of this is due to the imprecisely known vicissitudes of the martian surface, there are many other cases where some parameter in the program or limits placed on monitored quantities results in a "fault" being declared by the computer controller, and the rover ceases to rove further and simply awaits further instructions from Earth.

Likewise, it is impractical to know everything about a system as complicated as a human being, even if we have a whole-DNA sequence at our disposal. Thus, the "practice" of medicine is, to some extent, just that. It is likewise impractically expensive to assume that any one individual will be studied so precisely that their particular body systems will ever be fully understood, even just with respect to the more likely malfunctions or diseases.

4.4.23   UNDERSTANDING THE SYSTEM

One strategy for obtaining a better understanding of a certain biological System, rather than concentrating just on details of various subsystems, is to endeavor to identify and describe separate sub-functions of individual subsystems and investigate how they could affect overall system response. This is challenging, yet it is how engineering subsystems are actually designed. That is, in engineering design, it is the ultimate function, or functions, that is the goal, while leaving open how they are to be achieved.

Biological systems are restricted to using components that have been used previously with time-tested functions, not new functions. With modification of gene duplications and other methods of building upon proven beginnings, a new function is born. But they are not optimized. In engineering, there is extensive analysis of the robustness and failure-proofing of circuits and critical components to assure success *before* they are built and incorporated. Each subsystem is understood at both its lowest, most detailed level, and its highest, fully functional level. Robustness, reliability, resistance to environmental perturbations, and mean-time-to-failure are all understood at a sufficient level of detail and confidence to assure overall performance. In biology, systems and subsystems develop by constrained random walk. Even non-optimum developmental paths may have been followed but frozen into the contemporary design because of past history. Understanding a total system requires evaluation of each subsystem's performance both alone and in combination with variations of other subsystems. Allowing each subsystem's performance to vary, in response to purposeful perturbations ("stress testing"), environmental vagaries or to internal flaws or failures, is key to understanding the potential for success, failure, or non-optimum performance of any given overall system.

4.4.24 Payloads

Some systems accommodate payloads of various types. A "payload" is generally something not needed for the system's primary function, but incorporated as a convenience, typically for transportation to a target. It is in this sense that a rover and its carrier are considered the "payload" on top of the rocket which is launching it into space and then sending it on its way to Mars.

Likewise, the Perseverance rover carried a helicopter as a payload, which it deployed to the surface almost immediately after landing. The helicopter does return to the vicinity of the rover, but it is not required for the rover to be able to operate and fulfill its objectives. In the future mission to retrieve the samples acquired by the Perseverance rover, advanced versions of the helicopter will be used as a backup for retrieval of deployed samples for delivery to the return rocket landed nearby.

As Perseverance acquires samples to be returned to Earth, they will eventually become the "payload" of the rocket which will launch them into space, after installing them into a sphere that will be deployed and become the OS (Orbiting Sphere). Subsequently, a special orbiter vehicle provided by ESA (European Space Agency) will retrieve the OS and place it into an SRC (Sample Return Capsule), which will re-enter the Earth's atmosphere and land the samples safely. The OS thus becomes the "payload" of the SRC. The various payload-vehicle interfaces are often one of the more challenging engineering designs because they typically must involve both the mother vehicle and its design team as well as the designers of the payload itself.

Science instruments are also often referred to as a payload. However, only in a few cases are such instruments actually deployed to the surface and become independent (e.g., seismometers left on the moon by the Apollo astronauts). We group such instruments under our Sensors category, because they inevitably contain very advanced sensing devices, although a typical instrument is a stand-alone "black box" that is a whole system unto itself, including dedicated computer, power supplies, signal-conditioning electronics, a structural case, and often mechanisms.

For reproductive humans, their most important payloads are their gametes. Without healthy gametes, there can be no zygote, and hence no offspring. But gametes are not their only contribution to the success of perpetuating the group. In the terrestrial animal world, parenting is typically necessary to nurture and protect the offspring until they have grown and matured sufficiently to tend for themselves. For this, even members of the group who do not themselves participate directly in producing offspring often contribute their time, skills, and efforts that enable the group to survive, prosper, and successfully procreate. Likewise, engineers create rovers, and obtaining the necessary support to allow them to produce the next rover depends to a major extent on the success or failure of the previous rover.

4.4.25   DISCIPLINE FOR ENGINEERING DEVELOPMENT

When the system is highly complex and concomitantly very expensive, a rigorously systematic approach is warranted. The general standard is to authorize a series of phases for development. Phase A is the first initial determination of the higher-level requirements for the system and a review of which parts and subsystems of the system have a known history of application. This "heritage" assessment provides an important gauge of how challenging the system design and implementation will be, in terms of probable mission success as well as whether cost overruns may be likely.

Phase B provides additional detail, as well as development of any challenging components and any critical tests that need to be performed to provide confidence that the needed performances can be achieved. These are generally performed at the subsystem or even lower level. Occasionally, a test article approximating the final system may be fabricated for certain critical tests. Upon successful achievement of a satisfactory Phase B, the sponsoring agency may certify a full acceptance of the system and its mission and proceed onto the next phase.

Phase C is generally intended for detailed designs of the bulk of the items, especially those of some criticality. Additional reviews are conducted by the agency and outside consultants to pinpoint any areas needing additional attention, and to provide confidence in the design sufficient to authorize final fabrication of all the parts, boxes, and subsystems needed for the final product.

Phase D is the assembly and testing of the final system itself, including exposures to critical environments such as thermal cycles, vacuum, and Mars atmosphere. Although it might be thought that exposure to simulated martian dust would be important to verify seals and thermal controls, this type of testing is considered to be potentially detrimental because it is so difficult to clean all parts again, and the vibration and shock of launch by a rocket could inadvertently force any residual particulates into sensitive areas. In general, "Test as you Fly" is the mantra, but some aspects of the flight simply cannot be satisfactorily tested on Earth. In these cases, theoretical analyses, modeling, and partial testing must receive greater attention than is typically done.

Phase E is the mission itself. This is often capped at a certain length of time or achievement of pre-agreed goals. As appropriate, once the approved Phase E is complete, the project may be able to justify one or more Extended Missions.

These separate phases provide a step-by-step methodology by which to judge whether a project's system development is on course or needs remedial action (which may range anywhere from suggested or mandated changes to outright cancellation of the project).

4.5   ARE THERE FUNCTIONS IN THE HUMAN BODY ANALOGOUS TO ENGINEERING FUNCTIONS?

Is it possible to parse the physiologist's designations of human subsystems into the engineer's breakout? This is a critical question, because if we can demonstrate concordances between a mechanical rover and the ultimate complexity in biological manifestations, i.e., the intelligent human being, this encourages the application to all systems in between.

Most subsystems for a rover do have close analogies to many living systems, including us humans. For us, our mechanisms subsystem includes the muscles that power our movements, analogous to the many motors on the rover. Both rover and humans need fundamental structure – the bones and strong tissue for us, and the metals and advanced composites that are the "skeleton" of a rover or, for that matter, an engineering system such as an automobile or airplane. We humans use our brain organs as our built-in computational capability and memory bank. And we use our ears, tongue, and vocal cords to communicate with one another via sound waves, as well as our muscles to convey information via touch or body language. Of course, these days we also communicate via e-mail and websites, using our fingers to activate keys to feed information into a computer and thence out into networks of digital communications.

Even bacteria have similar types of functions with an actin-like cytoskeletal "structure" and motility via flagella and pili mechanisms. They have computer-like processing via positive and negative chemo- and phototaxis responses. Communication occurs within the cell, including the translation and transport of information via RNA molecules from codon sequences in the DNA to the protein manufacturing function in the ribosomes. Communication with the outside bacterial community is also possible, as evidenced by quorum sensing responses and possibly other phenomena.

4.5.1   Classical System Description of Humans by Physiologists

The human body is seemingly infinitely more complex than a single human cell, given that it has fluid and nerve networks, as well as a complex set of immune responses to ward off invasions by bacteria, viruses, protozoans, and fungi. The study of human physiology has long utilized a systems approach by dividing the body into several different "systems" (which we would consider "subsystems," since the overall System is the human individual itself – however, we will recognize the physiologist's terminology of "systems" for human subsystems in the discussion below).

Viewpoints differ somewhat on how to break things down. Thus, the key human "systems" might be considered to be as small as six, as shown in Figure 4.7.

But a much more complete breakdown can be at least 12 systems, as shown in Figure 4.8.

The general purpose of each system is compiled in Table 4.2. A code number is assigned to each human system, and the 80 or so organs in the human body are

Human Body Organ Systems

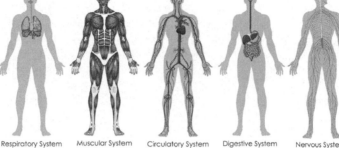

Skeletal System   Respiratory System   Muscular System   Circulatory System   Digestive System   Nervous System

FIGURE 4.7 Simplified classification of human subsystems. (From Shutterstock, with permission.)

HUMAN BODY ORGAN SYSTEMS

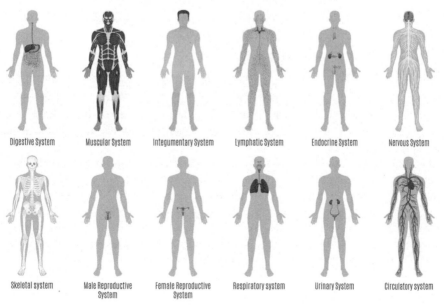

Digestive System   Muscular System   Integumentary System   Lymphatic System   Endocrine System   Nervous System

Skeletal system   Male Reproductive System   Female Reproductive System   Respiratory system   Urinary System   Circulatory system

FIGURE 4.8 A more extensive set of subsystems cited by physiologists for human beings, and generally referred to as the human body organ systems. (From Shutterstock, with permission.)

TABLE 4.2

Systems of the Human Body

Code	Human System	Key Functions	Key Organs
H-1	Digestive system	Absorption of nutrients from foods	Mouth, stomach, liver, intestines, rectum, anus
H-2	Respiratory system	Import O_2, export CO_2, Filter, Smell	Nose, lungs, pharynx, larynx, trachea, bronchus
H-3	Cardiovascular system	Circulate blood throughout body	Heart, blood vessels (arteries, capillaries, veins)
H-4	Nervous system	Control other body systems	Brain, spinal cord, afferent and efferent nerves
H-5	Endocrine system	Release signaling hormones	Glands: thyroid, pituitary, adrenal, pancreas, sex
H-6	Skeletal system	Support structure; protect organs	Bones, joints, cartilage, ligaments, tendons
H-7	Muscular system	Connect bones; enable movement	Voluntary muscle, smooth muscle (organs)
H-8	Urinary system	Export wastes; regulate fluid	Kidney, bladder, ureter, urethra
H-9	Reproductive, male	Male gametes: spermatozoa	Prostate, testes, penis, scrotum, deferens
H-10	Reproductive, female	Produce eggs, milk; nourish fetus	Mammaries, ovaries, uterus, vagina, fallopians
H-11	Immune system	Filter fluids; host white blood cells	Lymph vessels/nodes, marrow, spleen, thymus
H-12	Integumentary system	Protect exterior; evaporative cool	Skin, hair, nails, glands (incl. sweat), receptors

Source: Adopted with simplifications from Ashwell (2016), Hall and Hall (2021).

correlated with these systems. Here we have a challenge, because these 12 systems are not necessarily obviously analogous to our generic engineering system block diagram of Figure 4.6.

These physiologist-defined systems can be correlated with our engineering subsystems breakdown in some cases by selecting where there are multiple contributions. As a preview, a correspondence between the physiologist's "Systems" and the subsystems in the engineer's block diagram (Figure 4.6) is shown in the traceability matrix of Table 4.3.

The derivations of these correspondences can be considered on a case-by-case basis for each of the engineering subsystems, as follows.

4.5.2 POWER AND ENERGY

Metabolic usage of the daily intake of food energy generally proceeds via chemical reactions to convert this energy into molecules of ATP (Figure 3.3), the principal

TABLE 4.3

Traceability Matrix between Physiologist's "Human Systems" (H-n) and the Functions of the Generic Engineering Block Diagram (see Figure 4.6)

Engineering Generic Subsystems	Traceability	Human Systems
G-1	Power and energy	H-1, H-2, H-3
G-2	Control and data	H-4, H-5, H-7
G-3	Sensors	H-4, H-5
G-4	Structures and mechanisms	H-6, H-7
G-5	Import-export-storage	H-1, H-2, H-3, H-8
G-6	Communications	H-2, H-4, H-7
G-7	Internal environment	all
G-8	Mobility	H-7
G-9	Navigation	II-4
G-10	Reproduction	H-9, H-10
G-11	Defense	H-11, H-12
G-12	Interconnects	H-3, H-4, H-11
G-13	Boundary	H-12

molecule for storing and transferring energy in cells of all organisms. In animals, like we humans, the ATP is manufactured from the energy obtained by digestion of food and reacting it with the oxygen emitted by plants into the atmosphere. In contrast, the synthesis of ATP in plants is driven by the energy from capture of photons of light, i.e., photosynthesis.

Thus, the engineer's essential subsystem function of Power is accomplished for a human organism by chemically reacting the oxygen we breathe in with every breath of air, with organic molecules dissolved in our bloodstream, especially glucose. The human systems that contribute to energy production are therefore multiple: the digestive system processes raw food into smaller particles and converts the raw ingredients into smaller molecules, with a significant amount becoming glucose or other sugars to react with oxygen in air brought in by the respiratory system to produce ATP and NADH (Figure 4.2). The reaction actually occurs in the bloodstream, where CO_2 is exchanged with O_2 complexed with hemoglobin in red blood cells, and therefore is dependent upon a functioning cardiovascular system.

A diet of 2,500 food calories for a human per day is equivalent to about the same energy as the 125 W of electrical power from the nuclear battery to operate the rover. We humans, however, consume only slightly less power from the food we eat when we are totally quiescent, and only slightly more power when moderately active.

Although food energy is quickly depleted in general, animals like us can also store chemical energy in the form of fats and glycogen for later use. Fat molecules produce twice the energy for the same weight as glycogen and can do so without using oxygen. Some animals can hibernate and greatly reduce this daily energy consumption, and thereby "winter-over" for a few months by consuming their internal "battery" of these additional organic chemicals without seriously affecting their overall health.

4.5.3   CONTROL AND DATA

Perhaps the next most critical subsystem is the controls that must be placed on its activities and productive use of the energy that has been acquired. For a human, there are roughly three time scales of responses to stimuli. The fastest is involuntary movements due to detection of an abnormality or threat, such as dropping a very hot item or thrashing in response to a painful physical attack. The next fastest to these reflex actions is when the so-called "autonomic nervous system" controls various involuntary activities such as heart rate and breathing rate. The slowest responses are, of course, those mediated by conscious decision-making, typically by performing some trade-off analysis to determine a course of action.

The first example occurs when a sensory neuron directly stimulates a motor neuron in the spinal cord, bypassing the brain and triggering a reflex action at signaling speeds of up to 100 m/s. The latter two examples are conducted in appropriate portions of the brain, such as the hypothalamus for modulating respiration rate and the cerebral cortex for intelligent decision-making. The brain contains around a dozen discrete regions for controlling various activities, including the processing of various sensory inputs and subsequent activation of various other systems. Much of this is still bewildering to scientists, although through a variety of evidence, such as cases of specific brain damage or the monitoring of metabolic activity levels during various functional states using special MRI and PET-scan techniques, it is possible to help identify which areas of the brain specialize in which activities.

Certain brain segments also serve as storehouses for memory. The short-term memory, which often begins to degrade first with advanced age, is stored in the prefrontal cortex. Unconscious memory, which can be especially useful for skills such as athletic prowess in sports, or speed-typing on a keyboard, or proficient playing of a musical instrument, is stored in the cerebellum and the so-called basal ganglia.

Facts and information about events are stored in the hippocampus and neocortex, with help in emphasizing importance, signaled by the amygdala. Outputs from various physiologic sensors can also augment the strength of such memories, especially if they trigger one of the various emotional responses (joy, sadness, anger, embarrassment, surprise, disgust, fear, confusion, excitement, desire, contempt, etc.).

All these data sets are hugely important for human function and success in performance and decision-driven activities. Errors in memory or incorrect interpretation of a string of facts can lead to self-defeating actions by the human or other higher organism. Consolidation of time-driven "experiences" of past outcomes can lead to wisdom, and consequential wise decision-making.

Controls in humans are often mediated chemically, rather than by electrically driven neurons. Signaling molecules can be released by various "glands," which constitute the so-called Endocrine System. These "hormone" chemical communicators are very powerful in their ability to elicit physiologic responses, and complex in their interactions and sensitivities. As a result, major changes in human health can result from minor malfunctions of gland outputs. Some hormones are stimulatory while others are inhibitory. Overall, they can have enormous influence apart from neuronal-driven stimulation. Melatonin production induces sleep and can be inhibited by blue light from computer screens. Cortisol hormone signals stress, which triggers heightened preparation for action in dangerous situations. Irregularities in insulin activity cause diabetes. Thyroid hormones regulate metabolism, affecting the digestion of food and thereby the energy level of the organism. Endocrine gland releases of hormones are therefore a major complementary set of controls in addition to those regulated by the nervous system.

4.5.4 SENSORS

In the rack-up of human systems, the item of sensors is explicitly included in the "nervous system." Physiologists refer to sensory organs as "receptors." It is common to group several key ones into the so-called "five senses" (sight, hearing, smell, taste, and touch). In reality, animals are equipped with a much larger variety of sensors. For example, our blood pressure, pH, as well as concentrations of O_2, CO_2, and salts in our bloodstream are all monitored by a variety of chemoreceptive sensor organs ("carotid bodies").

Our immune system also uses molecular-signature receptors to detect the presence of bacteria. These are so sensitive and diverse, that they can respond to identify different bacterial species.

We also have "vestibular receptors" in our inner ear that detect head motion, body orientation (with respect to the gravity vector), and accelerations. This information can be used for the Navigation function and performance of athletic activities or fight-or-flight responses.

Sensory receptors can also have various degrees of sensitivity to amplitude as well as to spectral content. For example, some humans are "color blind" to one degree or another, and it is common for hearing sensitivity to higher frequency sound to degrade with age faster than sensitivity to lower frequencies.

4.5.5 STRUCTURES AND MECHANISMS

The primary structure is provided by the skeletal bones, some 200 of them in the typical human body. There are about three times this number of muscles to control the relative orientations of many of these bones, thereby creating numerous "mechanisms" and enabling everything from grasping to jumping to running to lifting to skiing, etc. These mechanisms also enable facial and body language, as well as exquisite verbalization for communication.

4.5.6 IMPORT-EXPORT-STORAGE

The digestive and respiratory systems are obviously major players in these functions, providing import of food nutrients, H_2O, and O_2. They also export wastes, such as feces, urine (especially excess nitrogen and salts), sweat, and exhaled breath (especially CO_2).

Energy reserves, such as provided by glycogen, fats, and ATP, can be stored in certain locations. Various enzymes detect needed nutrients in the environment and facilitate the Import of them across the exterior and interior boundaries. Others detect and sequester undesired excess concentrations of nutrients, as well as waste products and toxins, then transport them to the system boundary for Export.

Storage of food equivalents is of utmost importance, especially since it is most difficult to find, compared to water and, of course, air. Survival without key imports is sometimes quoted as the "three 3's": average survival without food is about 3 weeks; without water for 3 days; and without air, for 3 minutes before brain damage occurs. The human being is obviously in a highly precarious position with respect to many other animals when it comes to survival without importing our needed ingredients.

Likewise, we must export various things. Urination normally occurs several times per day and cannot be safely stopped for more than a few days (but birds do not urinate at all – they instead excrete nitrogen as uric acid, in their feces).

We export H_2O via sweat glands in our skin as a mechanism to enable evaporative cooling when overheated. However, in 100% relative humidity, this mechanism is useless. For this reason, survival above only 86°F (31°C) is doubtful in saturated humidity, whereas survival in much, much higher temperatures is possible under extreme dry conditions, as long as replenishment of drinking water and key electrolytes is available.

Export also occurs at the molecular level. When a cell is damaged by any of numerous factors, it typically will voluntarily self-destruct itself by a functional mode known as apoptosis (programmed cell death). Several other types of cells, especially macrophages, engulf and consume not only these cells but also invaders that the immune system has targeted for destruction.

4.5.7 COMMUNICATIONS

Communications for the human is multi-faceted and widely distributed. Apart from facial expressions, gestures, and other body language using our muscular system to communicate with our fellow beings, we have verbal communications of various types, ranging from small talk, to lectures, to warning cries, to singing, laughing, crying, and whistling. We also communicate intellectually through written language and emotionally by playing musical instruments.

Internal communication *within* the human body is yet more complex. The nervous system is a subsystem that includes the extensive network of nerve cells, concentrated in the brain but extending into the spinal cord and branching out to form the peripheral nervous system. This system operates at high speed, with fraction-of-a-second transmission of signals from sensors and similar times to trigger selected muscles. Some of these reactions are processed by the brain before being acted upon, but others are "involuntary" movements needed to protect the organism when time is of the essence.

Other methods of communicating within the human body involve passing of hormones via the bloodstream, the chemical signals for which are generated by a variety of organs and acted upon by a variety of other organs. This chemical signaling is used not only by animal organisms, but also by some plants and even the fungi. These signals are not only slower, but in some cases less specific in the responses they produce. Yet, the operability and health of the human body is intimately tied to the power of hormones to orchestrate critical longer-term response functions, ranging from nourishment to reproduction.

4.5.8 INTERNAL ENVIRONMENT

For humans, thermal control is aided not only by clothing but also by various internal functions which restrict or dilate blood vessels, as well as long-term acclimatization via accumulation or elimination of insulating fats.

Biological systems have a range of tolerance to thermal energy. Different bacterial species and strains have different optimum temperatures for growth and reproduction. Some are optimized for very low temperature, the psychrophiles (Oren, 2019). The hyperthermophiles (Oren, 2019), on the other hand, are optimized for extremely high temperatures by employing designs for enzymes that are unusually stable under high-temperature conditions (including the Cas9 enzyme that makes possible the powerful CRISPR gene-editing system). Individual biological cells are too small, in general, to sustain a significant thermal gradient against the strong conduction of heat by water or aqueous media.

But in larger, multicellular organisms, thermal gradients across sensitive organs can have detrimental effects. And for homeothermic organisms, such as birds and mammals, the maintenance of an internal temperature within a very narrow range is necessary not only for optimum performance of the system but for survival itself. In general, higher temperatures enable faster chemical reactions which therefore increase the speed and energetics of the organism until some limit is reached beyond which their internal metabolism becomes stressed and they perform with lesser speed instead of higher speed.

Other functions of "Internal Environment" include such activities as maintaining certain levels within specific ranges. In organisms, this function is termed "homeostasis." It can include maintenance of pH, osmolality, and the internal concentrations of key chemical nutrients such as the CHNOPS elements, as well as Na, K, glucose, CO_2, O_2, and even transition trace elements for synthesis of metalloenzymes. Homeostasis mechanisms also downregulate concentrations of potential toxins and either prevent their import or facilitate their export. When blood sugar levels are too high, the pancreas will release insulin; when blood sugars are low, it excretes a different hormone which causes the liver to supply glycogen to the bloodstream and inhibit the liver from storing glucose, to maintain homeostasis.

The overall method of achieving homeostasis is by the various sensors (receptors) stimulating hormones which counter the particular property that has gone outside predetermined limits. Feedback loops cause either increases or decreases in particular states, until the desired state is back within those limits.

4.5.9 MOBILITY

The ability to translocate the center of mass of an organism is distinct from the rooted nature of the higher plants. Humans can, of course, walk about or climb in order to change that location either sideways or up and down. This is especially similar to a rover's function, with a major exception. That exception is that the nearly universal technique for primates is the use of all appendages – both arms and legs – to achieve translocation. In contrast, the rovers implemented to date use wheels. Although legged rovers have been designed and tested, they are intrinsically much more complicated and typically heavier than steerable, independently powered wheels. Legs do have some advantages, such as ability for movement in highly complex terrain, such as a rough lava field, or to climb steep slopes, or to avoid rolling hazards such as deep sand or sharp rocks. Likewise, many primates are extraordinarily mobile in rough terrain, including steeply sloped scarps. They can negotiate trees with aplomb. Some primates even have a useable fifth appendage that enhances mobility, namely, their tail, especially if strong and prehensile.

4.5.10 NAVIGATION

As pointed out previously, various sensors support navigation, including the stereoscopic vision enabling depth perception in order to create a three-dimensional model of the environment. Brainpower provides pre-planning as well as reconstruction of previous traverse patterns. Many other techniques have been used by humans to determine where they are and where they have been, from the use of stored memories to the recognition of landmarks and positions of the sun or stars. Many of these techniques have been similarly employed by rovers, although current technologies are more sophisticated and more precise. Nonetheless, navigation has been essential for humans, both in exploring new territories and simply making efficient treks in well-known terrain.

4.5.11 REPRODUCTION

For human reproduction, it is 9 months to birth, but then ~10–15 more years to reach reproductive age (menarche for females and spermarche for males), but up to 25 total years to reach maximum physical performance and often much longer to attain optimum maturity in knowledge, analytical skills, and judgment.

Obviously, the topic of how human reproduction is accomplished is a complex and detailed one, and is quite unique from the method by which rovers are reproduced. Suffice it so point out that the widespread technique of sexual reproduction, i.e., by two different contributors, has the advantage of bringing together two proven sets of genes that may be substantially different in some respects and thus offer the possibility of some fortuitous cases of "the best of the best" being bestowed on a particular offspring. The gametes of the female and male are their primary "payloads" which will be united to bestow the resulting child with a truly unique complement but mostly redundant backup sets of genetic codes. The results are generally subject to happenstance, however, as compared to the changes to a next-generation rover which benefit from careful engineering judgment, design, prototyping, and prooftesting to assure each change is beneficial by one or more criteria before being fully incorporated into the new system.

4.5.12 DEFENSE

The human body contains elaborate defenses against attack by other organisms, including microscopic viruses, bacteria, protozoa, and fungi as well as macroscopic organisms of various size, ranging from insects to great cats.

The lymphatic or immune system in particular counters microscopic organisms which gain entry to tissues or the bloodstream. It does so by a combination of macrophages, B cells, T cells, antibody proteins, and other microscopic warriors which learn to recognize previous foes and respond with strong countermeasures. Even the thermal control system is sometimes brought into play in an effort to counter-attack the invaders by raising temperatures. This emphasizes that all systems are susceptible to upset by external forces or other systems, and may have built-in defenses that are not readily apparent until an attack causes responses that become measurable.

4.5.13 INTERCONNECTS

Within something as small as a typical prokaryotic cell, interconnecting networks are not necessary because the diffusion time for molecular transport is so very short because of the small submicron distances involved. Thus, energy, mass, and information can all be transferred quickly. For more complex cells, such as eukaryotes, there are transfers that are facilitated by structural pathways, such as protons transported across the mitochondrial membranes. And for multicellular systems, including the human one, there can be elaborate networks systems such as the neural networks of neurons connected by synapses; the cardiovascular network of arteries and veins; and the subsidiary network of lymph channels which eventually feed extracellular fluids into the bloodstream to distribute nutrients and eliminate waste products.

4.5.14 BOUNDARY

The human boundary is distinct, with its outer covering of skin plus some hair and nails. This boundary performs many functions, such as protecting against unwanted invaders and preventing leakage of blood, lymph, and conserving water and salts. Boundaries in systems should not be overlooked, because of their function in preserving survival and identity. They also provide essential control over what is kept or allowed inside, and what is kept excluded from the overall system.

Many other epithelial tissues also occur within the body, to segregate and protect the 80 or so key body organs. It is not uncommon for more than one layer of somewhat membranous tissue to be employed to provide protection for multiple ingredients or subdivisions in an organ.

4.5.15 SYNOPSIS

It is now apparent that our generic engineering block diagram seems to encompass the wide variety of functions that make up any highly capable system. Although the implementation of these functions can be quite different, depending on the environment, the functional needs, and the technologies available (e.g., biological versus engineering), the overall block functions (subsystems) are generically common. Any system being analyzed is intrinsically susceptible to deconvolution into interacting subsystem specialties that are shown in the generic block diagram and should be so-subjected to careful analysis to assure no stone is left unturned in seeking an understanding of "how it works."

5 Systems Chemistry

5.1 INTRODUCTION AND BACKGROUND

We provide here an introduction and a brief background on systems chemistry. We summarize the contents of some key papers on this subject, to facilitate the forthcoming presentation on the applications to prebiotic chemistry, chemical evolution, and the emergence of life.

But, before we do this, let us pose some elementary questions. Let us bring up the "mantra" of systems in general, which states that a system is composed of the parts which interact in a way to produce a new emerging feature, which is not contained in any of the individual parts. Let us start, instead of the chemical parts, with simpler parts, such as rocks. Let us imagine that the rocks are falling into the river gradually over time, and thus their numerical diversity is increasing. We have more and more rocks, but, still, they comprise just a pile of rocks. At some point these rocks will eventually make a dam in the river, which will be a new and emerging feature. This example comes from the considerations of the quantity-to-quality ("q-2-q") transition (Kolb, 2005, and the references therein) in which the quantity of the rocks generates a new quality, namely a dam. When the same considerations are applied to chemistry, the picture is more complicated, which is also discussed in the same paper. We present here a very brief account. Let us assume that we have "chemical parts" which are oxygen and hydrogen molecules. Upon combining them, they can react and produce a new "quality," namely water, a new chemical molecule which has different properties than its parts. This well-known case is easy to understand. While hydrogen, when mixed with oxygen, is flammable, water is not, as just one new property. However, this quantity-to-quality change has not occurred by a simple physical combination of molecules or their constituent atoms. Instead, the electrons that surround these were rearranged. This example of a chemical change is fundamentally more complex than the example of the pile of rocks which forms a dam. This chemical example makes a simple chemical system. But in the systems chemistry approach, we usually do not consider simple chemical reactions as systems. If we would, we would have literally millions of such systems which would just show the products forming from the reactants. Thus, although the condition of the system is fulfilled in any chemical reaction, namely that the product exhibits new features that are qualitatively different than those of the starting parts (reactants), we usually reserve the term systems chemistry for much more complex systems.

In Section 5.2, we give definitions and the scope of systems chemistry as proposed by the practitioners in this field of study since its inception.

5.2 DEFINITIONS AND THE SCOPE OF SYSTEMS CHEMISTRY

Systems chemistry is a newcomer among other fields which study complex systems, such as engineering, economics, computer science, biology, and ecology (Systems

DOI: 10.1201/9781003225874-5

chemistry: Wikipedia). The history of the development of systems chemistry reveals the growth of the field in fits and spurts (Strazewski, 2019a, 2019b). This is reflected in the proposed definitions of systems chemistry, which show a tension between the ambitious all-inclusive definitions and the narrower ones, which are more focused and realistic. Definitions that are futuristics are quite optimistic and are not readily achievable at this time but may be relevant at a later date due to the rapid development of this field.

The term "systems chemistry" was introduced in 2005 (Kindermann et al., 2005), in the paper titled "Systems chemistry: kinetic and computational analysis of a nearly exponential organic replicator." In the same year, the early definitions for systems chemistry were given (von Kiedrowski et al., 2010). These are cited in Table 5.1.

These definitions cover many areas, including supramolecular structures composed of many molecules. But they are focused mostly on the astrobiology goals of elucidating the origins of life, prebiotic chemistry, Darwinian chemical evolution, and the formation and composition of protocells.

In 2007 the European Union research network on systems chemistry came up with more definitions, which are summarized in Table 5.2.

Entry 2 from Table 5.2 states that the field is opened for chemistries of limited prebiotic plausibility. In this book we do not cover such chemical systems, since they are in general of limited or remote value for astrobiology at this time. The other definitions from Table 5.2 show a tendency to be a composite of many aspects of systems chemistry, so one has to examine them carefully and extract the aspect that is of immediate applicability to one's astrobiology project.

The process of building new and modified definitions continued and culminated in the launching of the *Journal of System Chemistry* in 2010 (published by Springer), with the scope which is listed in Table 5.3.

The entries from Table 5.3 give us a set of goals of systems chemistry in general. The goal from entry 4 is not easily related to astrobiology. The scope 10 calls for research from the conjunction of different fields of study which is at this time difficult to achieve.

TABLE 5.1
Early Definitions of Systems Chemistry, Proposed in 2005

1. A conjunction of supramolecular and prebiotic chemistry with theoretical biology and complex systems research addressing problems relating to the origins and synthesis of life
2. The bottom-up penchant of systems biology toward synthetic biology
3. Searching for a deeper understanding of structural and dynamic prerequisites leading to chemical self-replication and self-reproduction
4. The quest for the coupling of autocatalytic systems; the integration of metabolic, genetic, and membrane-forming subsystems into protocellular entities
5. The quest for the roots of Darwinian evolvability in chemical systems
6. The quest for chiral symmetry breaking and asymmetric autocatalysis in such systems

Source: Cited from von Kiedrowski et al. (2010).

TABLE 5.2
Selected Definitions of Systems Chemistry in 2007 as Proposed by the European Union Research Network on Systems Chemistry

1. Systems chemistry is seen as the joint effort of prebiotic and supramolecular chemistry assisted by computer science
2. The origin of life is seen as a major stimulus to organize research, but the field is open for chemistries of limited prebiotic plausibility
3. Subsystems may be classified as genetic, metabolic, or compartment-building
4. Systems chemistry studies the questions of asymmetric autocatalysis and chiral symmetry breaking
5. The key challenge of systems chemistry is to find the roots of Darwinian evolvability in chemical systems

Source: Modified from von Kiedrowski et al. (2010).

TABLE 5.3
Scope of the *Journal of Systems Chemistry* When It Launched in 2010 (From von Kiedrowski et al., 2010), as Shown by the Topics It Considers for Publishing

1. Experimental and theoretical studies of complex molecular networks
2. Catalytic and autocatalytic systems
3. Self-replicating and self-reproducing chemical systems
4. Dynamic combinatorial chemistry
5. Emergent phenomena in molecular networks
6. Information processing by chemical reactions
7. Bifurcation and chiral symmetry breaking
8. Bottom-up approaches to synthetic biology and chemical evolution
9. Research on chemical self-organization inspired by the problems of the origin and synthesis of life
10. Research from the conjunction of supramolecular, prebiotic, and biomimetic chemistry, theoretical biology, complex systems physics,and earth, planetary and space sciences with a center in chemistry

The *Journal of Systems Chemistry* was discontinued in 2015. It was relaunched in 2018 and published by a different publisher (M. Volkan Kisakürek, Switzerland). As of the writing of this book, this journal is still accepting submissions, on the subjects that are given in Table 5.4.

The entries in Table 5.4 are the same or similar to those in Table 5.3, except for the new entry of xenobiology. This is a new subfield of biology, with the objective to design forms of life with different biochemistry, including a different genetic code, than the life on Earth. This topic can be related to extraterrestrial life.

A further insight into the development of the goals of systems chemistry can be obtained from the descriptions of the Systems Chemistry Gordon Research Conferences (GRC) in 2018 and 2022 (see website: https://www.grc.org).

TABLE 5.4

The Scope of the *Journal of Systems Chemistry*, Which Was Relaunched in 2018 by the Natural & Life Science Publishers from Switzerland NLS, as Presented by the Research Topics it Considers for Publishing

1. Complex molecular networks
2. Catalytic, autocatalytic, self-replicating and self-reproducing systems
3. Dynamic combinatorial chemistry
4. Emergent phenomena in molecular networks
5. Information processing by chemical reactions
6. Bifurcation and chiral symmetry breaking
7. Bottom-up approaches to synthetic biology and chemical evolution
8. Chemical self-organization inspired by the origin and synthesis of life
9. Xenobiology
10. The conjunction of supramolecular, prebiotic, and biomimetic chemistry, theoretical biology, complex systems physics, and earth, planetary and space sciences centered in chemistry.

TABLE 5.5

Goals of the 2018 Gordon Research Conference on Systems Chemistry

1. Recognition that life is the product of complex systems of molecular reactions, connections, and interactions which give rise to a highly dynamic and functional whole
2. Design and preparation of functional biomimetic systems such as artificial cells and tissues
3. Design and study of complex systems, i.e., of dynamic, self-organized, multi-component chemical networks
4. New approaches for the construction and design of dynamic mesoscale materials
5. Study of supramolecular chemistry, origins of life, and far-from-equilibrium systems
6. Convergence of disciplines such as chemistry, physics, bio-engineering and structural biology on the central topic of complex chemical systems

Source: https://www.grc.org.

GRC's are at the forefront of science, and presentations at the conference are not made public. However, the objectives of the conference and the titles of the conference sessions and presentations are available on the GRC's website (https://www.grc.org). Much can be learned from these about systems chemistry developments. The first such conference was held in 2018 and was titled: "Systems chemistry from concepts to conception." In Table 5.5, we show some objectives of the conference, and in Table 5.6 the titles of selected sections.

Goals 1 and 4 appear most directly related to astrobiology, while goal 2 is not since it is not included in the astrobiology road map.

These titles are directly relevant to astrobiology, except perhaps for topic 5.

The next Systems Chemistry Conference occurred in the Summer of 2022. In Table 5.7, we present the description of the goals of this conference, and in Table 5.8, we list the titles of some sessions of this conference.

TABLE 5.6

Titles of Some Sessions of the Gordon Research Conference on Chemical Systems in 2018

1. Bottom-up construction of complex chemical systems
2. Alternative genetic systems
3. Compartmentalized and catalytic chemical networks
4. Nucleic acid systems chemistry
5. Dynamic functional materials
6. Energy dissipation in dynamic systems
7. Chemical reactivity far from equilibrium

Source: https://www.grc.org.

TABLE 5.7

Goals of the 2022 Gordon Research Conference on Systems Chemistry

1. Establish a theoretical and experimental framework for understanding, analysis, modeling, and design of complex systems that show life-like functions
2. Address the following topics: dissipative self-assembly, nonlinear dynamics, active materials, the origins and synthesis of life, controlled molecular movement and motility, and the engineering and modeling of complex systems

Source: https://www.grc.org.

TABLE 5.8

Titles of Some Sessions of the Gordon Research Conference on Chemical Systems in 2022

1. From self-assembly to life through modeling and experiment
2. Origins and synthesis of life
3. Fueled reaction networks and metabolic materials
4. Emergent catalysis, machines, and motility
5. Active and adaptive materials through molecular self-assembly
6. Active and adaptive materials based on nanoparticles or colloids
7. Biological systems chemistry
8. Systems chemistry in flow

Source: https://www.grc.org.

The titles of the presentations at the GRC on systems chemistry in 2018 and 2022 show the research areas in systems chemistry which are pursued by scientists at this time.

In conclusion, we have given an overview of the goals for systems chemistry and their development over time. Old goals are being updated, and new goals are continuously being added. Some of the systems chemistry goals are very broad, complex, and inclusive of quite diverse subgoals. Many of the goals of systems chemistry are applicable to astrobiology at some level: directly, indirectly, or remotely, but a few are not, as far as we can envision. Our approach is to keep focus on the systems chemistry which is relevant to astrobiology. This is shown in Section 5.3.

5.3   SYSTEMS CHEMISTRY AS RELEVANT FOR ASTROBIOLOGY

A tutorial review, "Systems chemistry" covers the basics of the systems chemistry field and covers pioneering work on this subject (Ludlow and Otto, 2008). An important aspect of their review is that it emphasizes a distinction between chemical networks that are under thermodynamic control versus those under kinetic control. This aspect is highly relevant to life, notably since life itself is an out-of-equilibrium system. Their review also provides insights into the organizational principles of molecular networks of autocatalytic and replicating molecules which are responsible for emergent properties, such as life.

Systems chemistry studies emergent properties of complex mixtures of interacting molecules, such as self-assembling systems and replicating molecules. Importantly, kinetically controlled molecular networks have greater relevance to biology than the thermodynamically controlled ones. The reason is that most biological systems operate far from equilibrium. Systems chemistry also looks at self-replicators, which are molecules that promote their own synthesis from a mixture of reactants. Self-replicators are based on RNA and peptides, among other molecular species. These systems can show complex behavior, such as symbiotic cooperation.

Ludlow and Otto also expand on the self-assembling systems. They start with the simple noncovalent interactions between molecules and give examples of how these can lead to the emergence of larger structures.

These authors state that the capability of chemists to design and create new molecules is now extending from isolated molecules to molecular networks. The latter complex systems need to be unraveled and linked to the origin of life where possible.

A more recent review, "The beginning of systems chemistry," points out that systems chemistry has its roots in research on autocatalytic self-replication of biological molecules, notably nucleic acids (Strazewski, 2019a, *Life*, p. 11). Another review, "Systems chemistry" (Ashkenasy et al., 2017), covers topics such as emergent properties, out-of-equilibrium self-assembly, but also challenges such as creation of concurrent formation-destruction systems, incorporation of feedback loops, continuous maintenance of a system that is far from equilibrium, and pushing replication chemistry away from equilibrium.

The article "Prebiotic systems chemistry: New perspectives for the origins of life" (Ruiz-Mirazo et al., 2014) provides an extensive background on prebiotic chemistry. They review prebiotic synthesis of biologically relevant molecules. Then they cover various aspects of prebiotic systems chemistry, such as emergence of self-assembly, autocatalytic networks, and networks of replicating species, among others.

In conclusion, we suggest what we believe is the most pragmatic approach to systems chemistry as relevant to astrobiology. We should start with the general definition that systems chemistry studies complex systems, and then determine what the complex system is. Depending on the system we study, we would then choose the subsystems which are appropriate for the specific case we study. For example, if we study prebiotic chemical reaction systems which are catalyzed by metals, we need to include a geochemical subsystem that contains metals. If the chemical reaction system proceeds in the dark, we do not need to worry about UV light as a source of energy, but need an alternative. Thus, our choice of subsystems is customizable. This is just one of the many examples we show throughout this book.

6 Prebiotic Chemistry

6.1 INTRODUCTION AND BACKGROUND

Prebiotic chemistry is the foundation of prebiotic systems chemistry. It deals with the synthesis of the "parts" of prebiotic chemistry systems, namely, synthesis of biologically relevant compounds. The most notable early success of prebiotic chemistry was the synthesis of amino acids under simulated prebiotic conditions (Miller, 1953). Prebiotic chemistry is currently reasonably well-developed and, in some instances, has reached a quite sophisticated level. From simple, plausible precursor molecules, it can yield the essential building blocks of life, such as amino acids, sugars, nucleic acid and their components, peptides, lipid-like compounds that mimic some features of membranes, as well as quite complex compounds that have potential for functioning in the primitive metabolism and genetic systems. These building blocks could lead to protocells. Could this process occur within a primordial soup? Its early version, shown in Figure 6.1, depicts the primordial soup as a system comprised of four chemical subsystems which lead to the emergence of the primitive cells.

We now turn to the subject matter of prebiotic chemistry by which the building blocks of life were made. This topic was extensively reviewed (e.g., Miller and Orgel, 1974; Pereto, 2019). It is outside the scope of this book to repeat such reviews.

However, the problem of the origin of life is not solved by synthesizing individual biological compounds. There is more to life than its components. Life is a system, which means that it is more than its parts, and that new properties emerge from the interaction of parts. Life requires a certain level of complexity, organization, interactions, networking, and communication between its components. Furthermore, life is a nonequilibrium system which maintains its status via complex interactions with the environment and by developing a way to store chemical energy which is needed for life's maintenance.

The unassisted prebiotic chemistry, such as that occurring naturally outside the laboratory, can yield numerous chemicals, tens of thousands or more, which, however, may present only a heap of chemicals, without prospects of evolving to the complex system such as life. A good example are chemicals that are found in the Murchison meteorite (Schmitt-Kopplin et al., 2010, 2015). Some of these chemicals are biologically relevant, such as amino acids or nucleobases. However, chemicals in Murchison exhibit only numerical and diversity complexity, but not organizational complexity. They do not interact among themselves in a way to produce novel functions and emergent properties which could lead to life. However, at least hypothetically, these chemicals may undergo further chemical evolution if they are placed in a different environment, which, for example, contains reactive chemicals, various catalysts such as minerals, metals, and/or clays, and displays favorable reaction conditions and useable energy sources.

DOI: 10.1201/9781003225874-6

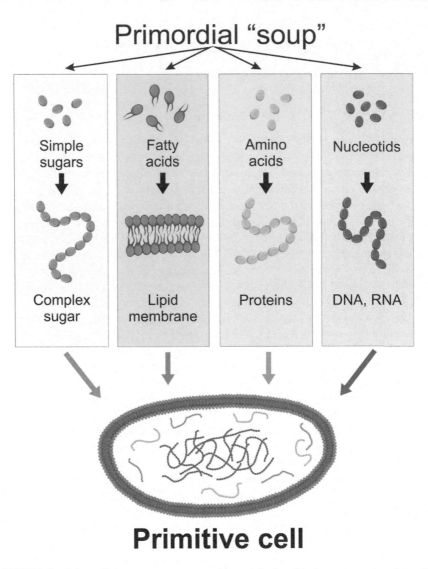

FIGURE 6.1 Primordial soup system: an early model of prebiotic systems chemistry that led to the primitive cells. (With permission from Shutterstock.)

In a contrasting example, a combination of simple prebiotically feasible chemicals may result in complex chemical features, such as autocatalysis, which emerge from such a mixture. We describe here one such case, based on the recent work by Wołos and coworkers (2020).

These authors first performed computer simulations of chemically feasible prebiotic reactions, starting from a small number of simple prebiotic molecules, namely, CH_4, NH_3, H_2O, HCN, N_2, and H_2S. Importantly, these are also molecules that are found in space (https://en.wikipedia.org/wiki/List_of_interstellar_and_circumstellar_molecules), and thus would have been presumably available for the chemistry of Murchison meteorite. The computer simulation by Wołos et al. (2020) predicted numerous products and networks of molecules, which contained all reported syntheses of biotic compounds from these starting materials, but also new, unreported synthetic routes. Several of these new synthetic pathways were validated experimentally. These include prebiotic syntheses of acetaldehyde, diglycine, and various acids, such as malic, fumaric, citric, and uric. In addition to these novel prebiotic syntheses, they discovered three types of chemical emergence. Firstly, they found that the molecules within the network can act as catalysts for subsequent reaction types. Secondly, they discovered the emergence of chemical systems comprised of self-regenerating cycles, which form within only a few synthetic generations. This type of chemical emergence is considered a critical step in building organizational complexity in prebiotic systems. The self-regenerative cycle of iminodiacetic acid was confirmed experimentally. Thirdly, they identified the network of chemical pathways to surfactants, both peptide-based and long-chain carboxylic acids. Surfactants are important components of prebiotic compartments. Computer simulations that were used were customized for the purpose of prebiotic organic syntheses. The authors developed these simulations by using 614 reaction rules which were grouped within 72 broader reaction classes, the knowledge of the underlying reaction mechanisms, and the feasible reaction conditions, all of which were reported in the literature. The reaction rules that were considered did not include stereochemistry, however. The calculations these authors performed are based on the Alchemy "Life" module, which is freely available (https://life.alchemy.net). Thus, the work by Wołos et al. is significant also since it showed that the computer simulation of prebiotic reactions networks is a quite useful tool for investigating prebiotic chemistry.

More recently, prebiotic chemistry researchers realized that prebiotic chemistry systems that led to life need to become competent to carry out and maintain the out-of-equilibrium status of the system, a feature that is necessary for life. This status is also an important requirement for the abiotic-to-biotic transition. It can be achieved in a system that is open to matter and energy exchange with the environment (e.g., Kompanichenko, 2017, 2020).

6.2   PREBIOTIC CHEMISTRY AS AN EXPERIMENTAL SCIENCE

Prebiotic chemistry is studied mostly experimentally, under simulated conditions of the presumed early Earth's atmospheric, geological, and hydrological environments. Chemical reactants are the elements, molecules, and chemical compounds that are thought to be available on early Earth. Various materials such as minerals, metals, and clays, which could have participated in prebiotic chemistry, typically as catalysts, should ideally be included in the experiments. As needed, simulations can utilize a variety of energy sources that were prebiotically feasible, such as UV light, electrical discharge in the atmosphere, and/or thermal energy from volcanos or hydrothermal vents, among others (e.g., Miller and Orgel, 1974; Miller, 1953; Kolb, 2019b, pp. 3–13). Sometimes it is difficult experimentally to include all these conditions, and the experiments are run in a somewhat simplified, reductionistic manner, namely excluding some of conditions that are assumed to be less important for the reaction output.

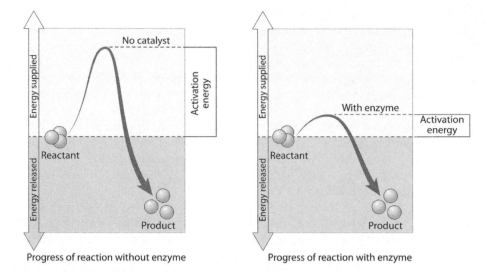

FIGURE 6.2   A comparison of a non-catalyzed and enzyme-catalyzed reaction. The reaction is faster if the activation energy barrier between the reactants and products is lower. (With permission from Shutterstock.)

Today's biological reactions occur with the help of enzyme catalysts. The purpose of a catalyst is to speed up chemical reactions by lowering the energy barrier between the reactants and products. This is shown schematically in Figure 6.2 in which a non-catalyzed reaction is compared with the enzyme-catalyzed one. Enzymes are very efficient and are often exquisitely reaction-specific, which evolved over a long period of time. In contrast, prebiotic reactions occurred presumably either without enzyme-like catalysts, or with some limited primordial catalysts such as minerals or metal ions. The primordial catalysts are in principle much less efficient than those from contemporary biological enzymes.

We do not cover here prebiotic syntheses of the common biologically relevant compounds, such as amino acids, sugars, nucleobases, lipid constituents, or the more complex compounds which were synthesized from these building blocks. Likewise, we do not cover the types of catalysts that may have been used in these syntheses. These topics are well-known and were thoroughly reviewed numerous times (see, e.g., Miller and Orgel, 1974, for the review of the early work, and Peretó, 2019, for the recent review). Instead, we focus on the prebiotic experimental designs for the syntheses of the building blocks necessary for the development of the early genetic systems, such as nucleosides and nucleotides. These syntheses are extremely challenging, and only recently significant progress has been made after years of repeated experiments which gave poor results. Progress was achieved by dropping the early experimental designs, which mimicked the biological pathways, and adopting new designs which abandoned such pathways. These novel syntheses often use powerful catalysts. This topic is covered in the next subsection.

6.2.1 PREBIOTIC SYNTHESES OF NUCLEOTIDES

The early prebiotic chemistry researchers assumed that the prebiotic syntheses of nucleic acids should follow the biological sequence by which these compounds were made *in vivo*. Thus, one should start from a nucleobase, such as adenine, and react it with the sugar ribose or deoxyribose, to form a nucleoside, and then phosphorylate the nucleoside to form the nucleotide. This method seemed hopeful, since the prebiotic syntheses of some nucleobases (from the oligomerization of HCN), sugars (from the formose reaction), and the phosphorylation processes were developed to some degree early on (e.g., Miller and Orgel, 1974). It was believed that the nucleotide synthesis could be accomplished by chemical "stitching" of the constituent parts. An example of a structure of a nucleotide with its constituent parts is shown in Figure 6.3. More such structures are readily available in any biochemistry textbook (e.g., Voet and Voet, 2011).

However, when the syntheses of nucleosides were attempted by the chemical stitching of nucleobases and sugars, many gave poor results in terms of yields. This was dubbed as "nucleosidation problem." This was discussed in depth in a recent paper by Yadav et al. (2020). These authors critically reviewed the reasons for the failure of the "stitching" method, some of which are intrinsic to the chemistry that was required, such as the reduced nucleophilicity of the nucleobases, their insolubility in water which necessitated heterogeneous reaction conditions, and high temperatures which then led to decomposition and side reactions. We do not dwell into this problem further since the chemical "stitching" approach was eventually abandoned.

Prebiotic syntheses of nucleosides and nucleotides were later developed, which are not based on the biosynthetic steps for these compounds. These novel approaches

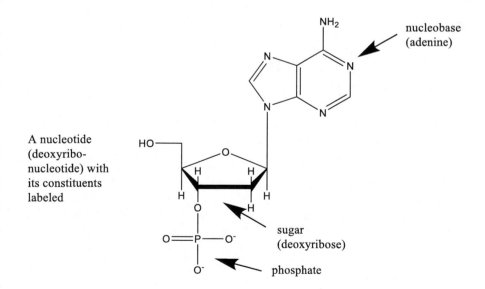

FIGURE 6.3 The structure of a nucleotide showing its constituent parts. (Modified and redrawn from Voet and Voet, 2011.)

were quite successful in producing the desired products, as shown by the Carell and Sutherland's research groups (Becker et al., 2018a, 2018b; e.g., Powner et al., 2011). Of particular importance are prebiotic pathways to RNA, since its formation is considered by many researchers as a key step in the emergence of life. Chemical principles of these new syntheses have been reviewed by Yadav et al. (2020).

Prebiotic chemical pathways to RNA were recently discussed (Strazewski, 2019c, pp. 235–264). The new pathways are quite complicated and require many assumptions. One key assumption is that such syntheses were very likely to have occurred on early Earth. Even if they did, it is not clear how these pathways can be sustained once the primordial chemical supply of key chemicals, such as HCN, dwindled. We may never know since the origin of life is a historic process. Still, these syntheses are highly important as models for how prebiotic syntheses could have occurred, and, upon further development, may be used as steps toward *ab initio* creation of life in the laboratory.

Another approach used in prebiotic chemistry is "top-to-bottom" one, as opposed to the previously described approach which is referred to as "bottom-to-top." The top-to-bottom approach attempts to reconstruct the origin of life starting from the most primitive life forms (the "top") and then moves backward toward the putative RNA world, until it finally reaches the "bottom," namely, primitive prebiotic chemical systems. New research tools, such as those in molecular biology, have facilitated this approach, and have contributed to recent advances in the understanding of RNA and the RNA world (e.g., Ruiz-Mirazo et al., 2014). Progress has been made in modeling the RNA world by studying RNA viral quasi-species (Padgett, 2012; Eigen, 2013). Coevolution of RNA and peptides is thought to be a part of the process for elucidating the origin of life. This was recently covered in detail (Strazewski, 2019d, pp. 409–420).

We keep our focus on the principles of prebiotic syntheses, and then consider which ones are amenable to the systems approach. Prebiotic chemistry systems are the foundation of autocatalysis and primitive metabolic cycles, as some examples.

In line with these goals, we cover in Section 6.2.2 the principles of multi-step linear vs. convergent organic chemical syntheses (e.g., Carey and Sundberg, 2007), in the prebiotic context.

6.2.2 Multi-step Linear and Convergent Syntheses in Regular and Prebiotic Chemistry

Syntheses of organic compounds may require multiple steps when compounds are complex and cannot be prepared in a single step. There are typically two ways to carry on the multi-step syntheses: linear and convergent (e.g., Carey and Sundberg, 2007). We start with the linear synthetic approach, in which the synthetic steps are consecutive, and thus occur in a sequential, "linear" fashion.

Traditionally, it was assumed that the multi-step prebiotic syntheses were linear syntheses. Therefore, this method has been predominantly used in prebiotic chemistry. However, it is known from organic syntheses that the multi-step linear syntheses generally give poor yields. Because of this, the linear multi-step prebiotic syntheses also give poor yields of the final products, even after accounting for other factors, such as possible intrinsic chemical reasons and lack of good catalysts.

In the multi-step convergent synthesis, which is typically used for the preparation of complex molecules, one part of the complex molecule is made in one reaction sequence, and the other part is synthesized in a separate pathway. Then the synthesized parts are brought together and reacted in the linear fashion to produce the final product. This synthetic method gives in principle a much better net yield than the purely linear one. However, it requires separate syntheses of the two parts of the molecule, and then a separate step in which these two parts are brought together and reacted to give the final product.

In the prebiotic convergent synthetic scenario, one compound would need to be made separately in one location, and the other one separately in another place, since the reaction conditions required for their syntheses are typically not the same. These synthetic steps must be coordinated in time and space. Otherwise, one of the compounds may react with something else or decompose before it has a chance to react with the other compound to give the desired product. Further, these two compounds need to be brought together by some means from the different locations, to be able to react to give the final product. Sutherland (2016) proposed prebiotic convergent syntheses in which the transport of the synthesized components from their initial synthetic locations to the final reaction site would occur by water streams and rivulets, and that the final reaction step would occur at the confluence of these.

Sutherland stated that it is not feasible that the prebiotic reaction networks occur in one pot, but that some degree of separation of the branches of each reaction network is necessary (Sutherland, 2016, 2017). Under such conditions, his overall synthetic scheme was successful. In 2019 Wu and Sutherland argued that the building blocks for life, notably ribonucleotides, were prefabricated from existing chemicals in the environment, rather than synthesized by emerging life (Sutherland, 2016, 2017; Wu and Sutherland, 2019). However, the probability of the chemical survival of the prefabricated chemical units during the transport from their initial synthetic location to the final reaction site, and the probability of these units finding each other in space and time, so that they can finally react, needs to also be taken into the account.

The systems approach to the linear single-pot syntheses and convergent syntheses which are based on the prefabricated units shows different requirements for these two methods. For example, chemical system in the single pot has a single boundary, that of the pot. The chemical system in the prefabricated units method has a boundary which includes several subsystems involved in this type of synthesis. They comprise the individual chemical systems in which the components are prefabricated, the rivers or rivulets transportation subsystem which carries the individual prefabricated units from their locations to the final reaction site, which is yet another subsystem, represented by the confluence of the rivers or rivulets. The formation of the final product is sensitive to the probability of survival of the individual prefabricated units prior to achieving their co-location, and the probability of arriving to the co-location at approximately the same time.

A single-pot synthesis may therefore actually have an overall more favorable probability of producing the desired end-product. Recently, Szostak and coworkers achieved one-pot synthesis which led to template-directed RNA copying (Zhang et al., 2022). Their results support the view that the one-pot synthetic method can be useful in synthesizing prebiotically relevant organic molecules. Other research has led to additional one-pot syntheses, for RNA nucleosides (Hud and Fialho, 2019) and for RNA nucleotides (Becker et al., 2019).

6.2.3 SYSTEMS CHEMISTRY APPROACH TOWARD BUILDING CHEMICAL NOVELTY IN PREBIOTIC CHEMICAL SYSTEMS

In this subsection, we explore the process of building chemical novelty in the prebiotic chemical systems by two methods. The first one is by moving the boundary of the system toward the environment, and the second one is by allowing the mass transfer from the environment into the system.

The traditional prebiotic organic syntheses are focused on the preparation of the compounds which are considered desired, in a sense that they were planned for and anticipated. These syntheses are typically not geared toward the discovery of novel products. When an unexpected product is formed alongside the desired product, it is usually considered undesired and is not examined further. This approach is not conducive for recognizing novel and potentially prebiotically useful compounds. The preconceived notion of what is an important prebiotic compound and what is not limits our thinking. It also artificially constrains the information about chemical diversity within the prebiotic chemical systems, by taking out of the consideration the "undesired" products. However, chemicals that are undesired at some point in time may become desired later. We describe here one such scenario.

Chemicals that are the starting materials that give the desired product become depleted, and the planned synthesis of the desired product is not feasible anymore since it is not sustainable. However, it is possible that our desired product now can be made from the undesired chemical products by a different synthetic process. This could be accomplished by a variation of Kauffman's "adjacent possible" reaction principle (Kauffman, 2016), in which a further exploration of the chemical reaction space is made. The undesired chemical products hit the boundary of the system, which is in contact with the environment, which provides new chemicals. The reactions at the boundary cannot be predicted ahead of time, since we do not know which chemicals would be present there. The reaction at the boundary occurs, giving new products, some of which may be our original desired products, or some novel products which are prebiotically relevant. When the synthesis of these ceases because of the depletion of the chemicals which are needed within the system, the chemical system explores a new boundary, which is further into the environment. This process creates an opportunity for novel and unanticipated reactions. This approach shows that defining the boundary of the chemical system should be flexible, and not fixed.

In conclusion, by considering the reactions at the boundary of our system, we open the horizons for exploration of chemical novelty in prebiotic systems. The experimental designs of the reactions at the boundary of the system which allows the exposure to the new chemicals from the environment are currently out of reach.

The systems approach is quite useful for upgrading a prebiotic chemistry experimental design. These experiments are typically performed in a reaction vessel, with starting materials and a solvent, and may be heated externally to speed up the reactions. In some cases, a UV light, spark discharge, thermal energy, or other prebiotically feasible energy sources are used. However, the reaction vessel model for a prebiotic reaction system is quite limited. We notice that the vessel is a thermodynamically closed system (see Figure 4.4), since we only allow the energy into the system but mass exchange with the environment is not allowed. Such a design limits diversity of chemical reactions and compounds that are formed. The environment, which comprises both atmospheric and geological subsystems, needs to be somehow included in our experimental design. For example, the geological subsystem contains various metals, minerals, and clays that may catalyze or otherwise influence the reaction. To simulate an open system that would allow the mass input from the environment, we can add to the reaction vessel some catalysts which mimic those that are present in the environment. Thus, we would bring a part of the environment to the reaction vessel, so to speak. A disincentive for this reaction design, however, is that we would probably end up with complex reaction mixtures, which resist analysis and become designated as "intractable." However, with the progress of the analytical tools, the analysis of complex reaction mixtures and the search for novel compounds may become feasible in the near future. Further, some methods of natural (unassisted) purification of complex reaction mixtures are possible. Examples include geochromatography (e.g., Wing and Bada, 1991), crystallization, sublimation, and surface adsorption (e.g., Islam and Powner, 2017), all of which make the study of complex mixtures more feasible than before.

In conclusion, we need to abandon the old thinking, which is limiting the discovery of novel prebiotic pathways and compounds. Instead, we should consider moving the boundaries of the system into the environment and allowing mass input from the environment to the system. Although we may not be ready to fully implement these goals experimentally at this time, we may be able to do so in the future.

6.3   PREBIOTIC SYSTEMS CHEMISTRY

Highlights of the beginning of systems chemistry that is prebiotically relevant were captured by Strazewski in his review article titled "The beginning of systems chemistry" (Strazewski, *Life* 2019a, and the references cited therein). This paper features Orgel's work on enzyme-free template-directed nucleic acid chain elongation; von Kiedrowski's research on enzyme-free autocatalytic ligation of oligonucleotides; Ghadiri's work on autocatalytic ligation of activated peptide fragments; Szathmary's theoretical treatment on selection of the faster replicating population; and Eigen's contribution on error threshold on replication fidelity (Strazewski, 2019a, and the references cited therein).

However, not all chemical reactions that comprise some of these systems are prebiotically feasible. An example is Orgel's use of activated groups which allow for enzyme-free reactions, but synthesis of these groups is not workable under prebiotic conditions. This problem was recently remedied by using methyl isocyanide (MeNC), which is a prebiotic compound, as a potential activating agent. MeNC is the result of a prebiotic systems chemistry approach which combines atmospheric, inorganic, and organic chemistries (Bonfio et al., 2020). This compound drives the simultaneous activation of nucleotides and peptides in an aqueous solution. Its prebiotic use was extended further to the activation of amino acids, peptides, and nucleotides within vesicles that were formed under the prebiotic conditions, and which were capable of sustaining these activation reactions (Bonfio et al., 2020).

Prebiotic systems chemistry in general was advanced and made prebiotically more feasible by novel approaches that were pioneered and pursued notably by Sutherland and his group (e.g., Patel et al., 2015; Sutherland, 2016, 2017; Powner and Sutherland, 2011; Powner et al., 2011; Ranjan and Sasselov, 2016; Ranjan et al., 2018; Islam and Powner 2017; Ritson et al., 2018; Sasselov et al., 2020; Bonfio et al., 2020). Sutherland and coworkers introduced cyanosulfidic chemical pathways, which can form the precursors to amino acids, ribonucleotides, and lipids, starting from HCN and some of its derivatives, hydrogen sulfide as the reductant, UV light, and Cu(I)-Cu(II) photo-redux cycling. Different applications of their system may also utilize Fe, P, Ca, and wet–dry cycles. Their work is chemically quite sophisticated. We suggest that the readers with the expertise in chemical synthesis consult the original papers for details. Here, we summarize one of the Sutherland's papers, "The origin of life – out of the blue" (Sutherland, 2016). Sutherland provides examples of prebiotic systems chemistry and points out the central role of HCN in this chemistry. The part of the title "out of the blue" means "unexpectedly, without warning," but also refers to HCN, which in German is called "Blausäure," which in English means "Blue acid." Both names imply HCN, since they refer to the acidic character of HCN and to its derivation from Prussian blue, which is a deep-blue pigment that is composed of complex iron cyanides. Sutherland considers that a minimal cell is comprised of integrated subsystems, which are informational, metabolic, catalytic, and compartment-forming ones. In addressing prebiotic synthesis of informational subsystems, Sutherland addresses first the problem that while HCN gives the nucleobase adenine, it also reacts with the aldehydes that are needed for the synthesis of sugars. He points out that the nucleobase and sugar could be synthesized separately and then brought together in space and time to react with each other. This would be the case of convergent synthesis, which we have discussed earlier in Section 6.2.1. Sutherland states that such a scenario could not be excluded, but the reaction of combining the two parts is difficult for thermodynamic and kinetic reasons. We have already brought up these difficulties, termed the "nucleosidation" problem, as discussed by Yadav et al. (2020), in Section 6.2.1. Thus, he has developed a better way to accomplish such a synthesis, which does not use the modular approach reminiscent of the existing biochemical pathway, but instead is a brand-new method, which draws on prebiotic chemical systems and which includes plausible geological, atmospheric, and meteorite-impact conditions. The latter may be especially important for providing the source of phosphate for prebiotic chemistry. Notably, the phosphide mineral schreibersite, $(Fe, Ni)_3P$ is promising and has been studied extensively by Pasek and his coworkers (e.g., Lang et al., 2019, and the references therein). Sutherland built his case for the systems approach gradually and with much detail. We show here selected highlights, which are focused on prebiotic systems chemistry. Sutherland pointed out that only two subsystems that comprise life, the informational and catalytic, are the same for all three kingdoms of life since they include the same molecules—RNA and proteins.

The third one, the compartment-forming lipids, is different. Bacterial and eukaryotic lipids are mostly diesters of one enantiomer of glycerol 1-phosphate, or its derivatives, with fatty acids. Thus, they are fatty acid-glycerol *esters*. In contrast, Archaeal lipids are mostly di-isoprenoid ethers of the opposite enantiomer of glycerol 1-phosphate, or its derivatives. Therefore, they are isoprenoid-glycerol *ethers*. This ester/ether difference is critical since ethers are more stable under some extremophilic conditions that Archaea experience.

Other researchers have also taken the systems chemistry approach to prebiotic chemistry. One example is Szostak (2009). Another is Powner, who originally worked with Sutherland, but now has his own research group. Islam and Powner's (2017) comprehensive review, titled "Prebiotic systems chemistry: Complexity overcoming cluster," provides the state of the art in this field. We present here some highlights. These authors provide details of the synthesis of RNA, lipid, and protein precursors by cyanosulfidic protometabolism, synthesis of activated ribonucleotides by bypassing ribose and nucleobases, and systems chemistry approach to prebiotic triose glycolysis, among other innovations. The cyanosulfidic origin of the Krebs cycle was experimentally supported by Ritson (2021). The Krebs cycle is central to metabolism and is shown in Figure 4.2.

Harrison and Lane (2018), however, advocate that life, thus biochemical pathways, should be used as a guide to prebiotic nucleotide synthesis. They offer detailed support for their view, such as a lack of resemblance of biological pathways to cyanosulfidic protometabolism, wet–dry cycles in UV-seared volcanic pools, the difficulty in extrapolating backward from LUCA (Last Universal Common Ancestor) to prebiotic chemistry, and the problem with the lack of accumulation of unstable reactive intermediates. They make the following statement:

> The demonstration that activated nucleotides can be formed from cyanide …proves it can be done, and it eliminates some of the mystique. But it does not prove this is the only way to do it. Life itself hints that this way was not the way it happened. If we want to understand the origin of life, we would be foolish to ignore life as a guide.

Also: "Perhaps the biggest problem is that the chemistry involved in these clever syntheses does not narrow the gap between prebiotic chemistry and biochemistry – it does not resemble extant biochemistry in terms of substrates, reaction pathways, catalysts or energy coupling." Harrison and Lane thus believe that prebiotic chemistry should be guided by that of the biological systems, which we understand quite well.

A fascinating example of a reaction system is provided by a recent study by Criado-Reyes et al. (2021). They have reinvestigated the famous experiment by Stanley Miller (1953), which we briefly review, before showing their work. Miller synthesized amino acids from H_2O, NH_3, CH_4, and H_2, by a spark discharge, in a specially designed apparatus. A simplified scheme of the Miller apparatus is shown in Figure 6.4.

Criado-Reyes et al. (2021) focused on the nature of the reaction flask when they repeated the Miller experiment, which always used borosilicate flasks. However, Criado-Reyes et al. questioned if the same results would be obtained if the flask was made of different material, and then answered this question experimentally. They used three different reaction vessels: (1) Teflon flask; (2) Teflon flask, inside which they placed some chips of borosilicate glass; and (3) borosilicate glass as a control. Their results showed that the nature of the reaction vessel plays an important role in

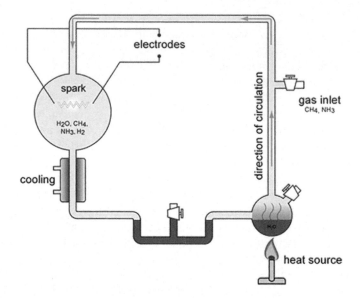

FIGURE 6.4 A simplified scheme of Miller apparatus for synthesis of amino acids by spark discharge from simple prebiotic chemicals. (From Wikimedia free depository.)

the reaction outcome. Thus, the molecular diversity is minimal in the Teflon reactor, it increases when chips of borosilicate glass are added, and it is the highest when borosilicate flask was used. In the latter case, the yield is also the highest.

Many advances in prebiotic chemistry have been made recently, but we have cited and presented here only selected key examples, notably those that are illustrative of prebiotic systems chemistry.

7 Systems Approach to Prebiotic Chemical Evolution That Led to Life

7.1 BACKGROUND ON CHEMICAL EVOLUTION

Chemical evolution is believed to be a part of the overall evolution of matter in the universe, as covered in detail in the classic book in this field, "Chemical Evolution, Origin of the Elements, Molecules and Living Systems" by Mason (1991).

From the systems point of view, chemical evolution means that a chemical mixture, comprising various organic and inorganic components, and with an energy input and the capability of converting this energy into a form that can be used by chemical reactions, evolved to give a chemical *system*. In this system, the parts of the mixture interact with each other to produce networks, autocatalysis, feedback loops, and other complex features, which have capability to lead to life. For life to function and be sustained, its chemical system must establish and maintain an out-of-equilibrium status, which requires capture of energy from the environment.

The evolutionary process, from the prebiotic chemical system to life, as exemplified by a primitive protocell, can be thought of as any other evolution in the sense that there is no specific direction in which the system will evolve. It all depends on the composition and interaction between the components of the system, the existence and interplay between various subsystems within the system, and the interaction with the environment including the mass and energy exchange. The system itself needs to be sustained, namely, chemicals need to be replenished, and energy must be delivered to the system. Some systems never make it. They are not sustained and they "die out." Their components may be taken up by other systems. However, some other systems do "make it," since they are better equipped for survival. For example, during the time when the supply of the essential chemicals from the environment dwindles or stops all together, these systems are able to produce the essential chemicals from other chemicals that are already present inside the system, or some other chemicals that are imported by the system from the environment and modified inside the system. Or they can enter a mode of stasis, with minimal needs, and "weather out" until the environment becomes more suitable.

A question arises about the nature of the initial composition of the chemical system which has a potential to evolve to life. Intuitively, one would think that the greater number and diversity of chemicals should be more favorable for such an evolution. However, the enormous diversity of chemicals that are found in the Murchison meteorite does not indicate that they have evolved to life in the past, since no remnants indicating past life have been found. Thus, the sheer number and diversity of

DOI: 10.1201/9781003225874-7

chemicals may be a necessary but not a sufficient condition for chemical evolution that led to life. One needs reactive chemicals, instead of inert ones. But then, if chemicals are too reactive, they are not selective and react with many other chemicals giving complex mixtures. The same applies to the chemical products that are formed, and which need to react further: they should be neither too reactive nor inert. Many more factors need to be considered, such as the presence of catalysts and energy sources. There are just too many unknowns that prevent us from providing solid predictions for the chemical evolution potential of a particular chemical mixture.

However, one promising lead for the successful chemical evolution of a prebiotic chemical system is its ability to establish autocatalytic cycles. Wołos et al. (2020) showed both theoretically and experimentally how a chemical mixture of prebiotically feasible compounds can produce autocatalytic cycles, as was discussed in some detail in Section 6.1.

7.2 CHEMICAL EVOLUTION THAT LED TO LIFE

One of the widest knowledge gaps in astrobiology is how chemical evolution led to life. Our knowledge is sparse, and many pieces of this puzzle are unknown. This is true for both the chemical pathways to life and processes by which the newly emerged life became sustainable. The transition between abiotic and biotic is especially poorly elucidated. We next elaborate on some of these problems.

Let us suppose that proto-life had some sort of RNA polymerizing/templating capability, some enzyme proficiency, and some primitive membrane functioning. Such primitive life may have been initially short-lived and would have decayed. There are three main reasons for this. The first reason could be that such a newly alive system would not yet have repair enzymes for RNA/DNA and would thus be prone to the information error catastrophe. This means that the fidelity of the replication would drop below a certain threshold of accuracy, which would render it useless. The second reason is that the proto-enzymes could not be coded reliably unless the properly working information was available, and thus would be prone to erratic behavior. Thus, a brief transition to life could quickly revert to the pre-life stage or, more likely, to a complete chemical decay. The third reason is a possible unreliable functioning of the primitive membrane, which needs to be more than a simple compartment of the prebiotic chemical system, but also the enabler of the specific in-and-out transport of the materials necessary for the functioning of the protocell. These problems can be consolidated if one adopts the view that chemical evolution that led to life occurred via a co-evolution of the prebiotic chemical subsystems, such as informational (RNA, DNA), enzymes, and membrane. This view is now beginning to prevail among researchers in the field (e.g., Bonfio et al., 2020).

However, the co-evolution between the subsystems must have some starting point in which the functioning of the system is at least minimal. There must be a sufficient complexity, organization, interaction between the subsystems, and an established and controlled exchange of matter and energy from the environment of the system for evolution to proceed. These requirements need to be further elucidated and hopefully confirmed. To accomplish this, we next look at the co-evolution of the prebiotic chemical subsystems.

We start from the definition of life number 24 from Section 4.2: "... a proposed definition of *life*: an open, coherent spacetime structure kept far from thermodynamic equilibrium by a flow of energy through it – a carbon-based system operating in a water-based medium with higher forms metabolizing oxygen" (Chaisson, 2003). This definition points out the spacetime (spatiotemporal) nature of life systems, as well as its far-from-equilibrium thermodynamic status. Relative to the spatiotemporal character of life, we believe that the prebiotic chemical evolution that led to pre-life and then to the transition zone (TZ) through which life emerged was likely dynamic in both space and time. Just like life itself, some products of chemical evolution were not everlasting. Some may have become "extinct" or may have been outcompeted and replaced with new ones. Some prebiotic chemistry processes may never have led to life; they just give a complex mixture of chemicals such as those found on the Murchison meteorite.

Next, we apply selected principles of systems thinking/analysis to the prebiotic chemical evolution that led first to pre-life and then life. We start with defining the system we wish to study. For example, we may look at the prebiotic chemical system which has a rudimentary metabolism, information, and membrane features. We may focus on the system that is considered pre-life, such as a protocell. Another system would be the transition zone that leads from proto-life to life. Chemical evolution would include all these systems, which, however, would not necessarily exist at the same place and the same time. For example, once life emerged, any chemical mixtures that are available would be consumed by life. Thus, prebiotic evolution would cease and would become an evolution of life itself.

The overall time span for chemical evolution that led to life is measured on an unknown time scale, which may range from years to millions of years. Chemical evolution may have been relatively fast or may have been punctuated, thus "fast, slow, fast, slow" and any permutations of these. This creates difficulties for our analysis. The environment also changes over time, influencing the evolution of life. For example, the atmosphere of the early Earth was initially considered to be reducing, which favored some chemical reactions. Later, additional research indicated that it was probably less reducing, or neutral, although the most recent analyses do address the early possible presence of reducing agents, such as H_2 and CH_4. These conditions would favor different types of chemical processes.

The interaction of the system with the environment includes the transfer of both mass and energy. This needs to be considered, as changes in the system occur. For example, chemicals in a system such as a small pond become depleted. Instead of declaring the dead end to the chemical evolution in this pond, we can enlarge the system to include a part of the environment which can provide fresh supplies. This could be a small stream that flows into the pond intermittently, depending on rainfall or a rise of the water table. Thus, some of our conclusions about the chemical evolution will be the result of our choices of the environment and the properties of the system we study. The example we provided is a dramatic one since it can lead to the conclusion of either a dead end or an active chemical evolution.

7.3   SURVIVABILITY, SUSTAINABILITY, AND LONGEVITY OF CHEMICALS DURING PREBIOTIC CHEMICAL EVOLUTION THAT LED TO LIFE

Our presentation in this subsection is based on the standard astrobiology knowledge which is available in various literature sources (e.g., Kolb, 2015, 2019a; Deamer, 2020; Dalai and Sahai, 2019; Cleaves, 2013; Chyba and Sagan, 1997; Baross, 2018; Baross et al., 2020; Zubay, 2020; Todd and Öberg, 2020; Wu and Sutherland, 2019; Szostak, 2017; Spitzer, 2019; Longstaff, 2015).

Chemical evolution likely took numerous turns and was influenced by various atmospheric, geological, and energy factors, all of which changed over time. While we can delineate some systems and subsystems which were part of the chemical evolution that led to life, most of this evolutionary process is of a historic nature and is thus unknown to us. Still, we can address specific difficulties in understanding chemical evolution, such as those that are related to the survivability, sustainability, and longevity of chemical systems that are involved, and the dependence of the organic syntheses on the reaction media and energy sources on prebiotic Earth. In this subsection, we address these difficulties in more detail.

Prebiotic chemicals that were present on the early Earth came from different sources. For example, some were brought to the Earth by meteorites, comets, and interstellar dust particles. This supply of material was not constant throughout the Earth's early history.

Some organic material was made by the reactions in the early Earth's atmosphere, on the surface, or in the aqueous environment, from chemicals that were presumably abundant at that time. Examples of the latter include HCN and various gases from volcanic emissions, such as H_2S. Not all the original material has survived long enough to produce sustainable chemical reactions. Some chemical material might have decomposed or polymerized into an unreactive chemical mixture. Once life evolved, critical organic material was continuously replenished within the organisms. Before this happened, the longevity and sustainability of organic material that was present on the early Earth was not assured. It is possible that some organic chemicals could have survived longer if they were complexed or otherwise stabilized by clays or various metals, among other options, and thus be available to react later when some new reaction partners appeared. If they were exposed to sunlight and other energy sources in the environment, they may have become highly polymerized material, with low reactivity and poor prospects of continuing chemical reactions. On the other hand, some reactions may have been promoted by the available energy sources to provide productive reaction sequences. For example, the power of UV light in producing pre-biologically relevant compounds has been demonstrated (Ranjan and Sasselov, 2016). We thus cannot make generalized statements based on the example of the Murchison meteorite.

The issue of survivability of the organic material needs to be examined in conjunction with its sustainability. The original sources for the organic materials that were provided by impacts of extraterrestrial materials diminished over a period of time. Destruction of key chemicals would give a bleak prospect for chemical evolution that led to life. If the chemicals were too reactive, they may not have been selective in their reactions and may have produced intractable mixtures. If they were less reactive, they would have been more selective, and thus would give cleaner reaction products. Even as we gain insight into survivability and sustainability of prebiotic chemicals on the early Earth, we still have the major problem of changing geochemical conditions over time. For example, phosphorus precursors have been proposed to come from meteorites. The low availability of phosphorus reflects negatively on the synthesis of ATP. This has led to the belief that prebiotic alternative sources of chemical energy existed, which were functioning early on, but were eventually taken over by ATP.

We also need to consider the longevity of a prebiotic chemical system, namely its duration throughout the chemical evolution processes which led to life. Longevity is survivability over a period of time. Thus, we need to take into consideration the time scale. The main steps of chemical evolution, such as the occurrence of individual chemical reactions, buildup of chemical complexity, and formation of proto-life systems, require different time scales. A simple prebiotic chemical reaction generally occurs much faster than the formation of the chemically advanced complex systems and especially proto-life systems, which may be measured on the geological time scale. During such a long timescale, the atmospheric, geological, and other factors that are important for chemical evolution, such as the already mentioned delivery of organics via comets and meteorites, may have changed.

One can consider a succession of the individual chemical reactions as a gradual process. When the chemical system becomes more complex, the gradual process may become punctuated, with changes which are introduced suddenly. This may occur when an individual chemical system is on a verge of becoming unsustainable, due to the lack of adequate chemical supply and/or adverse environmental changes. The system may survive by encompassing (fusing, overlapping with) another chemical system, which offers a related or alternative chemical supply or needed catalysts, and whose chemical constituents and reactions are less sensitive to the stress caused by the environmental changes. This fusion then enables the original chemical system to survive and to continue chemical evolution toward life. The lesson to be learned from this example of systems analysis is that we need to think about the chemical systems that are involved in chemical evolution as dynamic systems in time and space, which may fuse with other systems to survive, prosper, and evolve further.

7.4   CHEMICAL EVOLUTION IN DIFFERENT PREBIOTIC MEDIA AND ENERGY SOURCES

Chemical reactions in water, in superheated water, and in the solid state have been shown in the laboratory to give numerous products that are prebiotically relevant (Kolb, 2015, 2019e, pp. 331–340). Some of these products are quite complex. When such reactions occur in the natural setting, they have access to various catalysts, such as minerals, metals, and clays, which are present in the environment, and are expected to yield even more complex products. For example, hydrothermal systems generate unique diverse chemistry (Colin-Garcia et al., 2019). In addition, hydrothermal chemical systems have an additional benefit of being exposed to oscillating energy sources. The latter have been proposed to enable the out-of-equilibrium status of the chemicals system, which is an essential characteristic of life (Kompanichenko, 2017, 2020). This process is termed "Thermodynamic Inversion" (TI).

Let us examine the TI process via systems thinking/analysis. We focus on three subsystems that are present in the hydrothermal vent. These are the physical/ geological subsystem, the chemical subsystem within the vent, and the energy subsystem inside the vent which creates oscillatory conditions. When these three subsystems are coupled, they are proposed to create an out-of-equilibrium chemical system, which now exhibits a critical characteristic of life. The key to this model lies in the nonequilibrium thermodynamics and its relevance to the emergence of life, which is also discussed by Prigogine and coworkers (1972a, 1972b; Prigogine and Stengers, 1984; Prigogine, 1997), and by Chaisson (2003, 2015).

Without delving into all the details of the TI model, which are described in Kompanichenko's publications (2017, 2020), we point out some difficulties and knowledge gaps which need to be addressed before the TI model can be envisioned as fully functional. One main difficulty is the lack of the knowledge of the mechanisms by which the chemical subsystems within the hydrothermal vents are coupled to the available energy sources. In this respect, much of the other work on the origin of life and the abiotic-to-biotic (a-2-b) transition suffers from the same problem. Also, questions need to be posed about the survivability, sustainability, longevity, and evolvability of the out-of-equilibrium chemical system that is formed. How long is such chemical system going to survive in the hydrothermal vent? Is this process sustainable and how? How is the out-of-equilibrium status going to be maintained once the chemical system leaves the hydrothermal vent? To what extent does such a system have the capability to evolve? While we cannot answer these questions at this time, at least we can point to them and identify them as knowledge gaps. However, despite these knowledge gaps, the TI model involves an experimental setting in which an out-of-equilibrium system can be achieved. This setting is amenable to further experimentation.

7.5   EMERGENCE OF AUTOCATALYTIC CYCLES

Autocatalysis is the feature that is critical for the emergence of life and for functioning of life after it emerged. It is defined as the catalysis of a chemical reaction by one of its reaction products (Autocatalysis: Wikipedia). There is also a concept of autocatalytic sets of reactions. These represent a collection of reactions whose products catalyze enough of the reactions within the set to enable the set to catalyze its own production, and thus to become sustainable (Autocatalytic sets: Wikipedia).

The early work on autocatalytic cycles set the stage for further research (e.g., von Kiedrowski et al., 2010). The latter included chemical systems that were prebiotically more feasible. New types of autocatalysis were investigated. This topic is reviewed by Ruiz-Mirazo et al. (2014). In the recent work on the emergence of autocatalytic networks (Wołos et al., 2020), the authors generated these networks from the following chemicals that are commonly used as starting materials for prebiotic synthesis: CH_4, NH_3, H_2O, N_2, and H_2S. This work was discussed in Section 5.2.

The emergence of these autocatalytic cycles supports the idea of "metabolism first," which we now address. The topic of autocatalysis in the "metabolism-first" model was covered recently by Higgs (2021). Here we offer an approach that is the result of systems thinking.

The systems approach we suggest draws upon metabolism as it exists in living systems. There are two subsystems that comprise metabolism: catabolism and anabolism. Catabolism breaks down the food to produce energy in the form of ATP (for simplicity, we do not address other products here). Anabolism is responsible for the synthesis of biochemicals that are needed for cell functioning. This synthesis uses chemical energy in the form of ATP. The key to this systems approach is that the two subsystems, catabolism and anabolism, are linked via the energy flow in the form of ATP. These were briefly introduced earlier in Section 4.3 and are depicted in Figures 4.1 and 4.2.

In the present-day metabolic systems, various enzymes that are highly specific and are reliably made enable the reactions to proceed fast at room temperature. But, for the enzymes to be reliably synthesized, one needs an informational (genetic) system. We thus see how a systems approach in this case leads us to focus on co-evolution of the metabolic and informational subsystems, and not a merger of independent, fully functional subsystems. Metabolism cannot be reliably sustained or evolved by itself but needs to co-evolve with genetic systems.

Let us have a look at prebiotic reaction networks with an emphasis on synthesis (prebiotic anabolism). We start with the reaction network in which chemical products serve as catalysts for individual reactions. A qualitative look at this system, namely, starting materials and products, reveals the problem of specificity of catalysis. The early prebiotic reaction systems contained as catalysts smaller molecules and metals, but possibly also some larger enzyme-like compounds, perhaps of proteinoid nature. The latter, if they existed, did not have the means to be reliably reproduced, unless a primitive information system evolved concurrently. The early catalysis was probably not specific enough to maintain the reaction network. If primordial simple catalysis was not sufficient to facilitate chemical reactions within the proto-anabolic chemical system, then some other way to do so probably evolved. For example, chemical reactions could be facilitated by activation of chemical bonds. Such an activation could resemble that of the action of ATP. It seems that the catalytic function and the chemical activation need to be a part of the metabolic system. This complex process must be reliable. Without a proto-genetic system this would not be possible. Prebiotic chemical reaction systems need to be compartmentalized in some sort of membrane, which will allow selective mass and energy transport between a protocell and the environment.

We notice that autocatalysis is a necessary but not sufficient condition for metabolism. There are other factors that are important, such as the sustainability of the autocatalytic network and chemical energy which is required to perform chemical reactions inside the protocells.

In conclusion of this section, a systems approach may shed light on the metabolism-first model, by mimicking the biological systems organization, which means the inclusion of other subsystems, such as energy and genetic systems.

This discussion is related to that about the chemical evolution in Section 7.2, and elsewhere throughout this book. While enthusiasm exists for the metabolism-first, information system-first, and the compartment-first models, the systems approach argues for the co-evolution of these "firsts".

7.6   CHEMICAL EVOLUTION THAT LED TO LIFE: MACROBIONT AS A CRADLE FOR LIFE

In this section we describe in detail the macrobiont model for the pond hypothesis for the origin of life (Clark and Kolb, 2020). This macrobiont is a complex system that encompasses many environmental subsystems, such as geological, atmospheric, and hydrological. The complex interactions between these features culminate in the sequence of conditions that can favor an origin of life.

This model also pinpoints the transition to life, when a proto-macrobiont starts hosting live entities. More specifically, this model describes five stages of evolution within the macrobiont: prebiotic chemistry → molecular replicator → protocell → macrobiont cell → colonizer cell, thus bridging the gap between inanimate matter and a wider-scale biosphere. The "macrobiont cell" is envisioned as a more capable cell which evolves from the initial primitive protocell to be well-adapted to the macrobiont environment. It therefore multiplies extensively. The "colonizer cell" is postulated to be a derived version of the macrobiont cell line with capabilities suitable to a range of external environments which are more common than the macrobiont itself.

It may seem that such an overall primitive system is extraordinarily different from either an engineered rover or highly evolved and extraordinarily complex biological organisms. We shall investigate how systems thinking along the lines of identifying candidate subsystem processes analogous to those in our previous examples can expand this concept and expose potential interactions that otherwise might be missed or dismissed.

7.6.1   DESCRIPTION OF EXEMPLAR MACROBIONT

Many settings that have been considered as candidates for prebiotic chemical evolution which could lead to living organisms include: an organic-rich ocean; tidal flats; a warm little pond; suboceanic hydrothermal vents; volcanic hot springs; and combinations thereof. We choose as our exemplar the case of two or more separate ponds in a relatively dry environment, but which have intermittent connections to one another (Figure 7.1).

These ponds will contain solutes derived by dissolving gases from the atmosphere and by leaching of the soils and rocks the ponds are in contact with. These processes will provide many key components of importance to life as we know it, including the CHNOPS elements, namely, C and O from carbonates or organic molecules; H from water and organics; N from HCN or fixed nitrogen from the atmosphere, or nitrates; O from water, carbonates, organics; P from minerals; S from minerals and/or volcanic volatiles.

Discrete particulates, such as clays and organics, may play a key role. In response to the local gravity vector, buoyant particles and lower-density fluids will form a surface scum, while heavier constituents will form a bottom sludge. Extremely fine-grained particulates such as clays can remain in suspension or settle-out to form a top layer above the sludge. Soluble constituents will populate the bulk liquid, while evaporated products, such as salts, can form deposits around the edges of the liquid. These separations can be of significant value by providing a multiplicity of preferred

FIGURE 7.1 Multiple proximal ponds can be found in many geologic environments. (Left and middle, from Shutterstock with permission; right Photo Credit, B. C. Clark.)

locations for different types of prebiotic syntheses which can later be admixed with one another to further the processes of chemical evolution (Figure 7.2).

We now consider to what extent the previously identified subsystems may facilitate a systems approach toward functioning of a pond macrobiont system that could originate life. Let us consider how each subsystem in the generic block diagram of Figure 4.6 may be participating in such a macrobiont, at both the microscale and macroscale.

7.6.2 POWER AND ENERGY

All functioning systems require one or more sources of energy. Many studies of PCE (Prebiotic Chemical Evolution) have taken advantage of the fact that the early Earth was devoid of an atmospheric ozone shield, so that energetic solar ultraviolet radiation was freely available to drive certain chemical reactions. Portions of the UV spectrum can be quite deleterious, however, so it is important that some shielding or filtering also be available. Our pond can provide this with its surface scum and certain constituents (Ranjan and Sasselov, 2016) in the bulk solution. Even visible light wavelengths may be captured productively, as precursors to more elaborate photosynthesis processes.

Other sources of energy can be from chemical reactions between H_2O and the soil and rock minerals, especially for several possible redox reactions. Examples include a variety of sulfur oxidations of reduced sulfur when oxidizing power is available, or in the opposite direction when reductants such as atmospheric H_2 are

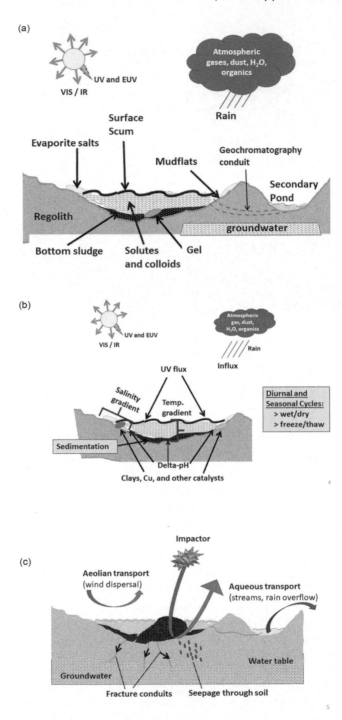

FIGURE 7.2 Depiction of a pond macrobiont. (a) Components of a dual-pond macrobiont; (b) dynamic aspects; (c) escape routes for the colonizer cell(s) to establish a biosphere. (Produced by co-author Clark; no permission needed.)

available. The acetyl CoA (Wood–Ljungdahl) pathway is the most primitive and simplest sequence for formation of important organic molecules via CO_2 fixation with H_2 (Ragsdale and Pierce, 2008).

Atmospheric products of UV photolysis are in chemical disequilibrium and could couple energy into PCE processes. Also, cycling of physical conditions can provide useful energy in promoting chemical reactions favorable to PCE. For example, wet-dry cycling and/or freeze-dry cycling are methods to promote dehydration reactions necessary to polymerize amino acids into proteinoids and nucleotides into oligomers of RNA or DNA (Ross and Deamer, 2016; Becker et al., 2018a; Deamer, 2019; Zhang et al., 2022).

7.6.3 CONTROL AND DATA

For the RNA World hypothesis, the ribozyme serves to control its own replication. As additional enzyme functions arise, controls are extended and refined.

The RNA nucleotide sequence is the code from which the various enzymatic activity arises and is also a storehouse of other encoded information. Thus, even before the protocell stage is reached, there should be data (information) stored in informational macromolecules which exert control on primary activities and molecular evolution.

7.6.4 SENSORS

When molecules change state due to irradiation by UV, they are serving as transducers. How this sensor-like behavior can be acted upon depends on the coupling with the overall chemical state of the macrobiont, which might also elicit very specific consequences on the microscale level that enable pathways toward the living state. Some molecules with catalytic attributes may be activated by solar photons, by changes in pH, or by other extrinsic factors.

7.6.5 STRUCTURES AND MECHANISMS

The shapes of the two ponds and the mutual connections between them can be of considerable consequence. For ponds with flat-lying shores and appropriate constituents such as certain clays and/or salts, not only can salty-rims form, but with repeated expansion and retreat of the shorelines, extensive mudcrack patterns of irregular polygons can form. These mudcrack arrays are created by the wet-dry cycling and hence can promote condensation polymerizations. They also provide local channels each with distinctively different UV and VIS irradiation profiles. In many cases, the topmost layers peel upwards from the polygon bodies to form shallow cups in which full-sunlight reactions can occur in highly concentrated solutions (Figure 7.3).

Through changes in water level, due to rainfall or groundwater recharge, access can change from semi-independent connection between ponds, to fully interacting. This passive mechanism of change adds considerable variability to local conditions and can mix components that evolved independently in the bulk ponds themselves with those that evolved in the crack patterns.

FIGURE 7.3 (a) Naukluft National Park, Namibia; (b) mudcracks on the banks of a channel; (c) mudpeels with salts. (From Shutterstock, with permission.)

Other structures relevant to the pond occur on the microscale. These include compartments (Deamer, 2000), such as vesicles (Deamer, 2017) or coacervates which enclose key molecules to achieve high concentrations of valuable constituents and protect them from deleterious compounds or more passive interference by waste products.

7.6.6 IMPORT-EXPORT-STORAGE

Importation can include, as noted previously, atmospheric gases that are useful for PCE. It has been hypothesized, based on atmospheric evolution models, that early in planetary history, reduced molecules such as HCN, CH_4, and H_2 may have been relatively abundant. Although their solubilities may be low in pond water, as they are used up in PCE reactions, the concentrations do not necessarily decrease since the atmosphere is an infinite source of reactants compared to the pond PCE products. Exportation of unwanted gaseous reaction products can also utilize the enormous atmospheric sink. Storage occurs in the form of equilibrium concentrations of those gases in the pond liquid, as governed by their solubility product constants.

Elements leaching during weathering of rocks and soils is a somewhat different case since the rates of weathering are typically slow compared to expectations for critical PCE reactions. However, many of these elements serve as natural catalysts and co-factors of proto-enzymes, so their concentrations need not be as high as needed for consumed substrates.

On a different time scale, the mudcrack fissures and peels can serve as temporary gravitational traps for volumes of liquids or solids whose composition becomes different from that of the bulk pond and could undergo additional chemical changes before being brought back into the pond by occasional flooding.

7.6.7 COMMUNICATIONS

The pattern of succession which we call "reproduction" can eventually develop inside the ponds. Down through the successions of progeny, information must be transmitted of the design plan for the next generation and must be done so in such a way that the error rate in copying is not so high that ultimately the critical information is eventually overwritten and lost. Thus, the capability for intergenerational communication of information is essential to ultimate success as the protocells evolve into more capable cells which, ultimately, are able to fend for themselves in the outside world.

These accuracies pertain not just to copying of the information itself for supplying the next generation with the master blueprint. There must also be fidelity in the copying of various portions of the information (genes) that are used in translating that information into the apparatus which fabricates the amino acid sequence to go into formation of the proteinoids which serve as enzymes, structural materials, membrane transfer agents, etc. Thus, the means of communication and interpreting the data transferred for conversion into essential products is a sine qua non for success that leads to creation of a biosphere.

7.6.8 INTERNAL ENVIRONMENT

Inside the ponds, the local environment is subject to the vicissitudes of weather and the chemical compositions of the external environments (soil, atmosphere). Inside the cells, there can be both passive and active processes that enable the interior to be different in beneficial ways from the bulk environment of the macrobiont itself. The PCE compartment's environment becomes more and more favorable as functioning protocells evolve to higher and higher levels of capabilities.

7.6.9 MOBILITY

A pond does not have the ability to translocate itself, of course, but on the other hand, it can change in size and shape depending on the weather (rainfall vs evaporation) and groundwater fluctuations. These changes in pond diameter are what provide the wet-dry cycling that is desirable for driving the formation of key biologically relevant polymers. It also provides expansion of the shore, including mudcrack arrays if the soils are appropriate. Whereas mobility by an individual organism is a way for the organism to experience a new environment, these expansion and contraction processes, plus the addition of new ingredients by runoff, are also a way to provide a new environment without translocation of a center of mass. In some cases, additional water can cause bridging between two or more nearby ponds that are not available in less wet times.

At the microscale, any ingredients such as particles, coacervates, and vesicles can be moved to different locations within the pond by various currents, as induced by winds or thermal gradients. Even in the absence of currents, there is diffusion and Brownian motion that will gradually re-distribute new constituents.

Transient weather phenomena can also perturb the ponds significantly. For example, dust devils are common in desert settings (both on Mars and Earth) and if such a whirlwind passes over a pond, it can cause stirring or sloshing to transport bulk material from surface to bottom, or even from one pond to the other. Microbursts can likewise create such effects. Ponds that are frozen-over are, however, temporarily protected from such perturbations.

7.6.10 NAVIGATION

Navigation within the pond is not under voluntary control, in general, unless some cells which formed have the equivalent of pili or flagella such that they can self-translocate, as well as have sensors to guide their direction or activity level.

7.6.11 REPRODUCTION

This feature is the most characteristic and fundamental of the properties we call life. It can occur first at the molecular replicator level, but eventually must evolve to the higher level of the fully functional cell in order to survive in a more hostile, outside world.

Although the macrobiont is the harbinger of life, it is not, itself, alive. The MB cannot reproduce itself. There may be clusters of ponds, and one pond may fill and flow over to create a nearby second pond. However, this is not reproduction guided by a blueprint. An analogy would be wildfires. Although such fires can grow and even "seed" secondary fires through their firebombs, they are not examples of life. Thus, although the macrobiont is not alive, it is an essential macroscale entity that enables the ultimate generation of living organisms at the microscale level.

Likewise, although the splitting of coacervates or membranous vesicles may mimic reproductive binary fission by microbes, these phenomena are not evidence of life unless their composition and progression is under central control by genetic information.

7.6.12 DEFENSE

Although a pond may eventually evaporate to dryness, as long as it remains as a topographic depression, it can be rejuvenated when conditions become more clement. If, on the other hand, the "empty" pond is covered over, e.g., by sediment or lava, it will cease to exist.

Likewise, if it freezes solid, the progression to advanced biological entities may temporarily be hindered. Or, if it becomes intruded by dust to become only a damp mud, its progress to an origin of life may be slowed.

However, a frozen surface may be a defense against various of these challenges and allow progression toward creation of living entities that eventually become colonizer cells.

At the microscale, compartmentalization can have positive effects by protecting key molecules, including those with genetic consequences, and avoidance of becoming ineffective from excessive dilution.

7.6.13 INTERCONNECTS

There are numerous modes of interconnectivity between different portions of the ponds. Mass movements of different constituents can occur by diffusion, advection, convection, wind stirring, and so forth. Thus, there is interchange by various physical processes driven by atmospheric and hydrological effects. The two ponds can be occasionally interconnected intimately, by a surface rivulet or subsurface water table. Connectivity can also involve chemical separation processes, such as by seepage which can include chromatographic separation effects, the so-called geochromatography (Wing and Bada, 1991).

7.6.14 BOUNDARY

Although the ponds themselves may seem to have a fixed boundary, the reality is that the shores are sometimes fully submerged and at other times devoid of water and hence not connected to the macrobiont itself. These fluctuating outer boundaries can be of critical importance because they provide the variable conditions which may facilitate the favorable concentration of ingredients as well as intermittent physical isolation of some portions where different PCE pathways may take place. In some ways they emulate the semipermeable membranes which, at the microscale, define the boundary of the individual cell.

There can be inner boundaries as well. Just as the rover is actually made up of numerous individual boxes of equipment, the MB will eventually be populated with individual discrete cells with their own micro-boundaries. Just as the human body is made up of several dozen discrete "organs" enclosed in epithelial tissues, the MB can have scum, sludge, suspended particulates, and possible aggregated clumps of complex material, each of which has discrete interfaces with the bulk fluid.

7.6.15 SUMMARY

From the examples discussed above, it is seen that even the origin of life itself can be considered as a large-scale system with most of the properties of an engineered system or a complex multi-cellular system. The origin of life is also an example of systems at the microscale which are embedded in a much larger system at the macroscale. The physical principles governing their operations can be rather different at these two scales. There also can be significant changes over time (evolution) at both scales, and for our purposes, this would seem to be essential just to facilitate the various changes that must occur to eventually achieve the creation of a microbe that can create, over time, a planetary biosphere.

8 Systems Approach to the Origin of Life, Including Abiotic-to-Biotic (a-2-b) Transition

8.1 SELECTED APPROACHES FOR STUDYING THE ORIGIN OF LIFE

Typical approaches by which scientists attempt to reconstruct the origin of life are bottom-to-top (*b-2-t*) and top-to-bottom (*t-2-b*) (e.g., Kolb, 2019b, pp. 3–13, and the references therein). Life is at the top, and the abiotic world that starts with primitive prebiotic chemistry is at the bottom. As chemical evolution gives rise to more complex chemical systems, emergent properties such as autocatalysis, self-replication, networks, and feedback loops eventually lead to pre-life. Somewhere in-between the pre-life and life is the transition zone (*TZ*) to life (Perry and Kolb, 2004). It is assumed that *b-2-t* and *t-2-b* approaches could give us convergent information about the origin of life.

The stages to pre-life, such as the RNA world, come from both the *b-2-t* and the *t-2-b* approaches. But this knowledge is fragmented and disconnected. The systems approach has the potential to produce a more integrated picture. As one example, for the transition to life, the pre-life system must already have in place a number of key subsystems that are needed for life, albeit they may be rudimentary and not always reliable.

Another approach for studying the origin of life is to break it down into two separate inquiries. The first inquiry looks at the origin of life as a historical process for which we are not able to find much concrete evidence, such as chemical fossils. The subject of the second inquiry is physicochemical processes in general by which life could have originated. This inquiry focuses on the intrinsic chemical reactivity, which could tell us which prebiotic reactions would have been feasible and most likely under the conditions on the early Earth, and on the potential chemical and thermodynamic processes, which could have led to the sustainable out-of-the equilibrium systems, which are characteristic for life. Understanding these may help us create an artificial life *ab initio*. If this approach succeeds, we could say that although we may never know precisely how life on the Earth came to be, at least we understand the principles of how it could have happened.

There are large knowledge gaps in these approaches. For example, the pre-life may have been characterized by some alternative genetic systems which were later taken over by the present one. We do not have support for the existence of such putative alternative genetic systems other than the belief that they may have existed due

DOI: 10.1201/9781003225874-8

to the fragility and chemical instability of some main components of the present genetic system. For example, ribose is not stable in solution even under neutral pH (Larralde et al., 1995), and RNA is labile in solution (RNA hydrolysis: Wikipedia; Ruiz-Mirazo et al., 2014).

Other approaches exist for studying the origin of life (e.g., Kolb, 2019b, pp. 3–13). The proponents of the "metabolism first" approach envision that the process starts with the development of a primitive metabolism. Others believe in a "membrane first" scenario, in which compartmentalization is the essential first step. The "genetics first" followers maintain that the formation of self-replicating polymers, such as RNA, is the critical first step to life. Peretó discusses the problems and controversies of these "firsts" approaches (Peretó, 2005). The most accepted belief is that the origin of life is linked to the RNA world, in which RNA is capable of self-replication, and possesses both genetic and enzymatic properties. As shown in our definition of life, we believe that life is a system, comprised of all of three "firsts," which are the subsystems of life. Even if one attempts to prioritize these "firsts," the fact is that we need all three of them for life. Again, systems thinking shows us the way to elucidate complex phenomena, not by just looking at the individual parts (or subsystems), but at their interactions which create the emergent properties which are not contained in any individual parts (or subsystems). This approach speaks for the co-evolution of all three "firsts" subsystems. Earlier, throughout this book, we have argued for this co-evolution from various points of view.

Life is a system that is out of equilibrium with its environment. Life is an orderly system, but its order is created at the expense of the environment. Thus, the more orderly life becomes, the more disorderly the environment is. Also, the energy that life uses for its various chemical reactions comes as an energy currency, namely ATP. In *t-2-b* approach, scientists try to mimic the approaches that current life takes to its energy harvesting, utilization, and maintenance of the out-of-equilibrium status. Progress has been made in elucidating some aspects of this problem, but a complete picture has not emerged.

The systems thinking/analysis approach is more successful in the *b-2-t* approach. We use here an example that was originally proposed and developed by Oparin (1924, 1938, 1966) and was recently reviewed and examined for its prebiotic feasibility (Kolb, 2016, 2019d, pp. 483–490). Oparin's model focused on self-organization and replication of primordial protocells. The protocells Oparin initially proposed did not contain DNA/RNA, structures of which were not yet discovered at that time, and thus he had no knowledge of them. Oparin conceived that the primordial protocells were coacervate based. Coacervates are colloid-based systems rich in organic macromolecules. They form in water, where they self-organize in either droplets or layers and exist as a *separate aqueous phase*. An artist rendition of coacervates is shown in Figure 8.1.

Oparin believed coacervate droplets to be protocells. These droplets can absorb organic molecules from the aqueous environment, which can react inside the coacervate to produce new chemical compounds. Coacervates can also absorb various catalysts, including the inorganic ones from the environment. This facilitates chemical reactions within the coacervate and leads to its growth. Due to the increase in their size, coacervates eventually become thermodynamically unstable and split into

FIGURE 8.1   Artist rendition of coacervates. (With permission from Shutterstock.)

daughter protocells. This splitting may also occur upon mechanical impact. The daughter protocells contain chemicals from the original coacervate droplet, whose chemical composition is thus preserved. This process would represent a primitive self-replication. While such a replicating system is not as reliable as the one that is guided by a genetic system, it may represent one of the options for early prebiotic evolution. Further, there is a possibility for selection among the primitive coacervate cells. These cells will compete for food from the environment. The faster-growing and faster-replicating cells will dominate the slower-replicating ones. This rate of the cell growth may be driven by the acquisition of catalysts from the environment, which is not expected to be uniform. Eventually, the food sources from the environment will become scarce and another level of competition would come into play. Coacervate cells that develop chemical pathways to make their own food would be selected over those which depend strictly on the environment for their food supply. We can see that this pathway of competition can be easily extended beyond what Oparin knew in the 1920s: coacervates that develop chemical ways to make RNA or its early equivalents will now reproduce genetically. The example of Oparin's coacervates can be extended, with some necessary modification, to other protocells. We would consider the following stages of the emergence of the protocell: Stage 1. Compartmentalization; Stage 2. Chemical reactions inside the protocells, which lead to their growth; Stage 3. Splitting of the cells into smaller daughter cells; Stage 4. Competition of cells for food; Stage 5. Chemical evolution inside the cells leads to the more competent metabolism and invention of genetic system.

These stages could be looked upon as a chemical system that is dynamic in time and within their environment. Coacervates and their nearby environment change.

Some coacervates may not survive, but their content may be recycled and taken up by the remaining coacervates. More in-depth discussion of Oparin's coacervates is offered by Kolb (2016, 2019d, pp. 483–490), including various shortcomings of the original model. We have provided here the core of his model, which we believe is more than of historical value.

Building on the Oparin's model, we can see that it could be applied also to the membrane-bound systems, such as vesicles, but the problems then exist with the in-and-out transport through the membrane. Still, these membrane-based models are being further developed to address this issue.

8.2 ABIOTIC-TO-BIOTIC (A-2-B) TRANSITION

8.2.1 AN OVERVIEW OF THE PROBLEM OF THE A-2-B TRANSITION

The abiotic-to biotic (*a-2-b*) transition is probably one of the widest knowledge gaps in astrobiology. Different hypotheses have been proposed for this transition. Examples include work by Spitzer (2019), Strazewski (2019d, pp. 409–420), Fry (2019a, pp. 109–124, 2019b, pp. 437–462), Kolb (2005, 2012, 2013, 2016), Perry and Kolb (2004), Krishnamurthy (2018), Walker (2015), and Smith et al. (2021), among others. Perry and Kolb (2004) proposed that this transition occurred via a transition zone (*TZ*), which possesses most but not all of the properties of alive systems.

By necessity, these hypotheses are intimately linked to and driven by the specific definition of life the authors adopt, and an understanding of the minimal necessary requirements for life which make life distinct from the abiotic systems.

Another knowledge gap is if the *a-2-b* transition is gradual or sudden. The views on this are split. The problem with either of these views is that we do not know the time scale for this transition. What seems sudden on one time scale may appear gradual on another. Perhaps a more realistic viewpoint is to consider how many complex steps must be invented and preserved on the pathway toward living entities. Even if we have knowledge about the time scale, the frequency of our observations may mislead us in believing that a gradual process is sudden. This is shown in Figure 8.2, which depicts the emergence of a butterfly from its caterpillar. If the time intervals for observations are limited to just the emergence of caterpillar and the butterfly, we would incorrectly conclude that this transition occurs suddenly, in just one step, and would miss all the complex transitional forms in-between.

In the case of the emergence of life on the Earth, we have evidence that life existed on the Earth ca. 3.7 billion years ago, which is the age of the oldest rocks with fossil evidence of life on the Earth. The most primitive forms of life may have emerged long before this time (https://en.wikipedia.org/wiki/History_of_life). However, we do not have knowledge about the time intervals, which would tell us if life's emergence was gradual or sudden and how many intermediary steps might have been involved. We even do not know if life emerged more than once.

The aspects of the *a-2-b* transition that include gradual vs sudden are discussed in, e.g., Kolb (2005), and will be reviewed in Section 8.3.

A lack of commitment by some researchers to stick to the term "life" as distinct from "not life," and instead use a fuzzy boundary between the two or even the term

FIGURE 8.2   A part of the life cycle of a butterfly that shows its emergence from its caterpillar. (With permission from Shutterstock.)

"not-life-yet," can cause confusion. We addressed this issue (Kolb, 2012), which will be summarized in Section 8.3.

The **a-2-b** transition is amenable to the systems approach since life emerges from the pre-life systems. As we have learned already in numerous other examples, emergence is one of the critical properties of any system. We shall briefly review some proposals about this transition in Section 8.3 and in Section 8.4, and we shall attempt to evaluate these proposals via systems analysis.

8.3   SELECTED PROPOSALS FOR THE ABIOTIC-TO-BIOTIC (A-2-B) TRANSITION

8.3.1   INTRODUCTION

Properties attributed most often to the chemical systems that are transitioning to life are: complex chemical behavior, such as self-organization and self-assembly; formation of autocatalytic networks and protometabolic cycles; and establishment of self-replicating systems (Kolb, 2019b, pp. 3–13; Padgett and Powell, 2012; Peretó, 2012; Ruiz-Mirazo et al., 2014). Peretó discussed the proposed stages in the origin of life, which comprise prebiotic, protobiological, and biological stages. Each stage consists of the additional sub-stages, such as the pre-RNA world (Peretó, 2005). Krishnamurthy (2017) proposed how a diverse pool of prebiotic building blocks could lead to a self-assembling system that is capable of chemical evolution from prebiotic chemistry to protobiology. His analysis included different classes of biomolecules that built the thioester world, protein world, metabolism world, lipid world, and the RNA world. More recently, the out-of-the-equilibrium thermodynamic status of life has received more attention and has been considered or implied to be a requirement for the chemical systems that are transitioning to life (e.g., Kompanichenko, 2017, 2020; Chen and de Vries, 2016; Chaisson, 2003, 2015). In addition, thermodynamic and kinetic factors are proposed to constitute the driving forces for the emergence of life (Pross, 2003, 2012).

We limit our presentation to selected exemplars of the *a-2-b* transition, which already use systems analysis or are promising for its application. In Section 8.3.2, we discuss the paper by Jeancolas et al. (2020) titled "Thresholds in origin of life scenarios." These proposed thresholds are intimately related to the emergence of life from abiotic matter and utilize a systems approach.

In Section 8.3.3, we discuss the quantity-to-quality (*q-2-q*) model for the *a-2-b* transition (Kolb, 2005). This model is abstract enough to be applicable to other cases, it is amenable to systems analysis, and it addresses the knowledge gap about the nature of the *a-2-b* transition regarding its sudden or gradual status.

In Section 8.3.4, we address the nature of the transition zone (*TZ*), "gray area," or "fuzzy" area between life and non-life, which necessitates a new way of thinking, namely dialetheism, and not Aristotelian logic. This example will illustrate how dialetheism leads to a new approach to defining systems.

8.3.2 THRESHOLDS IN THE EMERGENCE OF LIFE

First, we explain what the authors (Jeancolas et al., 2020) meant by threshold, and how the term threshold is related to the term emergence that we use. Next, we extract from their paper the essentials that are related to this chapter since the paper is very broad and comprehensive, as its title "Thresholds in origin of life scenarios" indicates.

According to these authors, thresholds define conditions of existence of particular states along the path from inanimate matter to life. The authors define a threshold as a major qualitative change in physicochemical system which is induced by minor changes in the system or its environment. The thresholds are found at different stages and levels of organization. They present different categories of prebiotic transitions. The coupling between the systems and their environments determines how thresholds can be crossed. In Table 8.1, selected examples of thresholds in the origins of life scenarios and the emergence of new qualities are shown.

The authors also present different types of transitions, listed in Table 8.2.

Much can be learned from the paper by Jeancolas et al. (2020), which we briefly summarized. Firstly, they did not leave a stone unturned when looking for transitions that are relevant to the origins of life, and they have taken a systems approach in many of their examples. Their work clarifies the nature of the transitions, but it also raises more questions. For example: are all these transitions necessary and sufficient for the primitive life to emerge, and if all of them need to occur simultaneously, how is this accomplished?

8.3.3 QUANTITY-TO-QUALITY (Q-2-Q) MODEL FOR THE ABIOTIC-TO-BIOTIC (A-2-B) TRANSITION

In this subsection, we re-examine portions of our paper "On the applicability of the principle of the quantity-to-quality transition to chemical evolution that led to life" (Kolb, 2005), which are amenable to the systems approach.

The idea that life emerged by a quantity-to-quality (*q-2-q*) transition of abiotic matter was originally proposed by dialectical materialists, who took the materialistic approach to Hegel's idealistic laws of logic. The philosophical background

TABLE 8.1

Selected Examples of Thresholds in the Origins of Life Scenarios and the Emergence of New Qualities

Selected Threshold Examples in the Origin of Life Scenarios:

1. Chirality symmetry breaking
2. Spontaneous polymerization
3. Self-assembly of compartments
4. Catalytic closure threshold
5. Error threshold

Examples of the Emergence of New Qualities from the above Threshold Examples, Shown as the System's State before and after the Threshold:

For 1) Before: racemic state; after: homochiral state

For 2) Before: solution of monomers; after: solution of polymers

For 3) Before: existence of free constituents; after: formation of molecular self-assembled compartments

For 4) Before: existence of polymers with few catalysts; after: formation of closed collective autocatalytic sets

For 5) Before: un-replicated polymers; after: polymers copied by template-based replication

Examples of Triggers for the above Transitions:

For 1) Enantiomeric excess, presence of circularly polarized light, autocatalysis

For 2) Ponds evaporation, freeze-thaw cycles, mineral surfaces

For 3) pH, temperature, salinity, wet-dry cycles

For 4) Number of catalysts and catalyzed reactions, spontaneous synthesis of diverse polymers

For 5) Copying error rate, genetic polymers' length

Source: Adapted with modifications and simplifications from Jeancolas et al. (2020).

TABLE 8.2

Different Types of Transitions in the Origins of Life Scenarios, with Brief Descriptions

Types of Transitions and Their Brief Descriptions:

1. Extrinsic transition: the acquisition of a new property is driven exclusively by the environment.
2. Intrinsic transition: the novel property is the result from changes in parameters of the system, while the environment remains constant.
3. Scaffolding transition: the environment changes only transiently but this change fixes a novel property within the system via a combination of the environmental change with a system-driven change. The environment acts as a *scaffold* since it transiently supports the emergence of a property that is then internalized by the system, namely, it becomes an intrinsic property of the system.
4. Symbiotic transition: two distinct systems aggregate into a new system, and thus acquire a novel property.

Source: Adapted with modifications and simplifications from Jeancolas et al. (2020).

for the q-2-q transition can be found in our paper (Kolb, 2005, and the references cited therein). Here we show a striking example of q-2-q which comes from Spirkin and Yakhot (1971). In our paper (Kolb, 2005) we offered a detailed analysis of their example, which now we examine via the systems approach.

Spirkin and Yakhot describe building a dam by a gradual addition of rocks to the riverbed. They stated:

> First a batch of rocks was thrown into the riverbed. There was no dam as yet. And there was still no dam after the second and third batches. However, a moment came when the number of rocks that had been thrown into the river was such that they began to have a radical effect on the flow of water. A few more rocks and the river was dammed …. While the quantitative changes were taking place within certain limits they did not seem to result in the formation of a new quality (in this case a dam). However, as soon as they reached a certain, definite quantitative limit, or measure, the changes began to produce visible qualitative effects.

The system analysis of this example is fruitful. We see that the system that is composed of the river and rocks initially does not have a purpose. However, as more rocks are thrown into the river, which can also be done naturally, since rocks can be dislodged from a larger rock along the riverside, a new quality, that of a dam, emerges. This would be a model for the evolutionary transition to a dam, and it would be sudden. The "purpose" in this case would be the functioning of a dam to stop the river flow. This function does not necessarily emerge, just as the evolution does not necessarily gives us life.

One of the conclusions of this example is that the *numerical quantity of the parts*, namely the number, and also the arrangement of the rocks, can lead to a new quality, that of a dam.

Next, we address another example from our paper (Kolb, 2005, and the references therein), which shows how *diversity of parts*, in addition to their number, can lead to the emergence of a system with new properties. The examples considered came also from the dialectic materialists. One example is the combination of two parts, oxygen, and hydrogen, to give water, which does not resemble the properties of the original parts and thus is a new quality. In our paper we ascribed this new quality to a new organization of the parts, atoms of oxygen and hydrogen, and their respective electrons. While chemical details of this new organization are offered in our original paper (Kolb, 2005), here we take a systems approach. The parts of the system that are diverse by their nature, thus intrinsically qualitatively different, can combine to give a new quality, whose nature (quality) is different than that of the parts from which it was made. In conclusion, both the number of the parts and their properties matter.

8.3.4 On the Nature of the Abiotic-to-Biotic (a-2-b) Transition

We have reviewed various ideas about the nature of the a-2-b transition and have proposed the existence of the transition zone (TZ) to life. We then hypothesized about its nature (Perry and Kolb, 2004). Further, we have published on the nature of the abiotic-to-biotic (a-2-b) transition (Kolb, 2010, 2012, 2016; Kolb and Liesch, 2008). Our papers examined questions such as if the a-2-b transition is sudden or gradual, and if the exact boundary between life and not-life exists. We have often addressed these questions with philosophical approaches, which are described in detail in our papers. Here, we offer a simplified version, which is most directly relevant to the systems approach, which we have not used in the original papers.

First, we address the question if the exact boundary between life and non-life exists. As we have shown throughout this book, the opinions are divided. One approach is based on Aristotelian logic, which allows for only "yes" and "no" answers and nothing in-between. Thus, it should be either life or non-life. However, another philosophical logical approach exists, that of dialetheism (two-way truth, in Greek), which allows for both answers to be true if the system that is considered is in a transition state (Priest, 2002, 2006, 2018). We give here a simple example that Priest used, namely that of a person exiting the room by passing through an open door. At some point in time and space, the person will be both inside and outside the room. This is easy to visualize, with one foot in each and the body part-way in both. This analogy can be applied to the *a-2-b* transition, in which the chemical subsystem in the *TZ* belongs to both abiotic and abiotic subsystems, and thus would be both alive and non-alive. In the Aristotelian logic we could only say that a system can be alive or not alive, and the nature of the *TZ* could not be handled as being both, as it can in dialetheism.

There are many ways to approach the question if the *a-2-b* transition is sudden or gradual. If we stick to dialetheism and then say that the time during which the transition occurred is very short, then we would have a sudden transition. If the time was long, the transition would be gradual. Since we do not know the time scale for this transition, this knowledge gap would cause the same dilemma as that which is depicted in the previously shown Figure 8.1 for the transition of a caterpillar to its butterfly. Does the caterpillar *suddenly* become a butterfly? Or is it more accurate to say that it is more accurate to say the transition was gradual. The accepted name for this hidden *TZ* is "pupa."

9 Systems Biology and Its Relevance to Prebiotic Chemical Systems That Led to Life

9.1 INTRODUCTION AND A BRIEF BACKGROUND

In this chapter, we introduce systems biology, which is a new scientific field, starting only about two decades ago (Systems Biology: Wikipedia; Ideker et al., 2001; Kitano, 2001, 2002; Chuang et al., 2010; Voit, 2020). Systems biology is much more developed than prebiotic systems chemistry. Thus, not all its principles of the former are applicable to the latter.

The overall goal of systems biology is the same as that of biology, namely, to understand how life functions. However, the systems biology focuses mainly on the interactions between the components of the biological system, and how these interactions bring about the function and behavior of these systems. This contrasts with the early approaches to biology which focused on the discovery, identification, and characterization of the components that are fundamental building blocks of life, which led to a rudimentary understanding of genes, proteins, and metabolites (Voit, 2020).

In addition, systems biology is characterized by many novel tools that come not only from biology, biochemistry, and biophysics, but also from mathematics, statistics, computing, and engineering (Voit, 2020). Modern systems biology analyzes a large quantity of data that is produced by the "omics," such as genomics, proteomics, and metabolomics, among others. The suffix "omics" means "a lot of it" or "all of it." A well-known example of omics is genomics. It studies not just one gene at a time, but most or all the genes at the same time (Voit, 2020). Omics studies result in a large amount of data, which creates the need for computational approaches to handle it. The systems biology thus has a computational branch, in addition to the traditional experimental one. The focus on the computational branch is reflected in one of the contemporary definitions of systems biology:

> Systems biology is the computational and mathematical analysis and modeling of complex biological systems. It is a biology-based interdisciplinary field of study that focuses on complex interactions within biological systems, using a holistic approach (holism instead of the more traditional reductionism) to biological research.
>
> *(https://en.wikipedia.org/wiki/Systems_biology)*

DOI: 10.1201/9781003225874-9

The article "Systems Biology: A brief overview" (Kitano, 2002) gives state-of-the-art coverage of the systems biology field at its inception in ca. 2000 (Kitano, 2001, 2002). Chuang et al. in 2010 published a paper titled "A Decade of Systems Biology," which testifies to rapid progress in this field. The authors applied a systems approach to the systems biology literature and found that the field has grown from a handful of publications published in 2001, to almost 2,000 in 2009. The rapid growth of the systems biology field is also reflected in the updates of the textbooks which cover this topic (Alon, 2006 first edition, 2020 second edition).

Chuang et al. (2010) reviewed the applications of systems biology to various complex problems, such as cancer, and the developments of systems biology software, neither of which are obviously applicable to prebiotic chemical systems. A more recent review titled "Systems biology primer: the basic methods and approaches" (Tavassoly et al., 2018) includes principles of the field, systems-level experimental analysis of cells, omics technologies, analysis of biological networks, experimental methods for systems biology, medical and therapeutic applications, and much more. Again, most material is not relevant to prebiotic chemical systems, but some are. We describe the relevant material in Section 9.2. In Section 9.3, we present the work which directly links systems biology and the origins of life (Ricard, 2010a, 2010b, 2013; Brunk and Marshall, 2021).

9.2   SYSTEMS BIOLOGY FEATURES THAT NEED TO BE ESTABLISHED IN THE PREBIOTIC CHEMICAL SYSTEMS TO ENABLE THE EMERGENCE OF LIFE

We start with some general aspects of systems biology which also apply to prebiotic chemical systems that lead to life and present them in a chronological order. Both the field of systems biology and prebiotic systems chemistry have significantly advanced during the past two decades, albeit not at the same pace.

Kitano wrote in his 2002 review of systems biology (Kitano, 2002): "To understand biology at the system level, we must examine the structure and dynamics of cellular and organismal function, rather than the characteristics of isolated parts of a cell or organism." Further:

> Identifying all the genes and proteins in an organism is like listing all the parts of an airplane. While such a list provides a catalog of the individual components, by itself it is not sufficient to understand the complexity underlying an engineering object.

And finally, "...to understand how a particular system functions, we must first examine how the individual components dynamically interact during operation." These citations clearly show the systems approach which we have followed in this book.

Kitano seeks a system-level understanding of a biological system by examining its key properties. These include system structures, which are often subsystems, and which include the network of gene interactions and biochemical pathways; system dynamics, which reveal how a system behaves over time; the control method, consisting of mechanisms that control the state of the cell and allow modulation to minimize malfunctions; and the design method, which enable us to understand how the system functions.

Let us examine how these could apply to the prebiotic chemical systems that can lead to life. System's structure and dynamics are definitely applicable. The control method had to be established late, since it is quite complex. As for the design method, the biological one is not fully applicable to the prebiotic system since the latter is much more primitive than the former and does not yet have all the capabilities and complexities of life.

Kitano further discusses *robustness* as an essential property of biological systems and compares it with the engineering systems. Out of this comparison, he finds that robustness is attained by system control, such as feedback; *redundancy*, in which multiple components with equivalent functions exist as backup; structural *stability*, which is achieved by intrinsic mechanisms to promote stability; and *modularity*, in which the subsystems are physically or functionally insulated so that failure in one module does not spread to other parts, which could lead to system-wide failure. Here we clearly see what needs to be built in prebiotic systems so that they can lead to life. In our opinion, redundancy is extremely important, and the mechanisms by which such redundancy can be achieved and maintained.

There is much more to be said about systems biology, but, for our purpose, this brief presentation may suffice to drive the message: Understanding life/biology as a system helps us set some specific features which prebiotic systems and pre-life must attain.

Systems biology advanced rapidly through the omics revolution. Much of the omics revolution is not directly relevant to the prebiotic chemistry systems, but some of its general principles are. We start by briefly covering just one of the omics, the genomics, as an exemplar of the omics, and then show how some of its principles can be applied to prebiotic chemical systems to formulate what we suggest can be named as a "*prebiotic chemomics.*" The latter would mean that instead of focusing on just one prebiotic reaction, we need to consider a myriad of possible reactions as they could have occurred under the actual prebiotic conditions.

Let us consider the so-called "messy chemistry" (Guttenberg et al., 2017) in which a prebiotic chemical system produces a large number and diversity of chemical compounds, which at this time can be analyzed only partially and with great difficulty, due to the limitations of the presently available analytical techniques. However, such a set of data, when obtained, would be open to the discovery mode. Various compounds that exist which were previously unanticipated could provide research leads into what chemicals are promising for creating higher-level interactions that could result in the systems' emerging properties.

We now expand on the hypothesis-driven research vs discovery-driven one in prebiotic chemistry. When doing the hypothesis-driven research, we mix selected prebiotic chemicals based on our hypothesis that they will yield desired products. When the latter are found we then validate this hypothesis. In the discovery mode our focus is not just on desired products, but on all the chemical products which are formed in the process, including undesired products. This is accomplished by mixing a variety of prebiotically feasible compounds without a prejudice about their structures. The discovery mode approach may reveal prebiotic synthetic potential we did not anticipate. This could be a fruitful approach in principle, but so far has not been fully realized. Let us remind ourselves of tens of thousands of chemical compounds that have been identified in the Murchison meteorite (Schmitt-Kopplin et al., 2010,

2015), and we still do not see much connection between them. While the discovery method vs. the hypothesis method is now a hallmark of success of systems biology, it has been much more difficult to implement in prebiotic systems chemistry. Thus, "prebiotic chemomics" is still in its infancy.

9.3   SYSTEMS BIOLOGY AND THE ORIGINS OF LIFE

In this section we review work by Ricard (2010a, 2010b; 2013, 2014a, 2010b) on the systems biology and the origins of life, in which he applies a mathematical approach to these topics. Ricard considers biochemical networks as possible ancestors of living systems, and examines factors such as reproduction, identity, and sensitivity to signals of biochemical networks (2010a), and (2010b) the nonequilibrium, self-organization, and evolution of the networks of catalyzed chemical reactions. In his book (2014b), which is a comprehensive review of his work, he addresses these and additional topics, such as the origin of information and communication in biological systems.

Ricard's work (2010a, 2010b; 2013, 2014a, 2014b) is an exemplar of the systems approach to the origins of life. He wrote: "The problem of the origins of life, or of the origins of novel functions, cannot be understood without having recourse to the concept of system" (Ricard, 2013). It is significant that he includes the origins of novel functions alongside with the origins of life. This is in line with the systems approach/analysis, in which any system must possess an emergent property.

Ricard further considers biochemical pathways of present-day cells. These pathways and their networks are the subject of much study in systems biology, but Ricard tentatively considers them as models of prebiotic systems. Further, he examines the biochemical networks as possible ancestors of living systems (Ricard, 2010a).

Before we continue with the presentation of Ricard's work, we provide his definition of life, which guides his research, and his consideration of what came first – RNA or proteins, which is one of the critical questions for the origin of life. In addition, we briefly address his mathematical method.

9.3.1   RICARD'S DEFINITION OF LIFE

According to Ricard (2010a), the living systems are characterized by the properties such as their ability to reproduce, their identity, and capacity to evolve. The biological events should be considered in the context of a history, and the behavior of living systems should be looked at in the context of a time-arrow. His verbatim definition is the definition number 42, in Section 4.2. However, in (2010a) he also elaborates on the various aspects of his definition, from which we select the following: "A living systems is an entity that should be considered a coherent whole possessing an identity specific for both its organization and functional properties. Today, this identity is defined from the structure of some macromolecules namely DNA and RNA."

9.3.2   SOME EXAMPLES OF RICARD'S WORK

Ricard explores the possibility that an encapsulated network of proteinoid-catalyzed chemical reactions can exhibit some types of spontaneous duplication and self-organization, which he has studied with mathematical models. One example is shown below.

Ricard's mathematical approach in studying simple chemical systems starts with consideration of probabilities of events within the system. The results then indicate if information is being generated, since probabilities and information are linked. This is illustrated in an example of a two-substrate proteinoid reaction which displays random binding of its substrates, A and B. The probability of the ligand A binding to the proteinoid is $p(A)$, but if the ligand B is already bound to the proteinoid, then the probability of binding A is $p(A|B)$. If $p(A)$ is larger than $p(A|B)$, this means that the interaction between A and B generates information. In general, the relationship between the information h and the probability p, for a material entity such as a molecule x_i is $h(x_i) = -\log p(x_i)$. For the above example we look at the difference between the two cases which gives the information for the system $i(A:B) = h(A) - h(A|B)$. If $i(A:B) < 0$, the interaction between A and B generates information.

9.3.3   A COMPREHENSIVE LIST OF TOPICS WITHIN THE GENERAL SUBJECT OF SYSTEM BIOLOGY AND THE ORIGINS OF LIFE BY RICARD

A list of topics studied by Ricard, which are within the general subject of systems biology and the origins of life, is summarized in Table 9.1.

TABLE 9.1

Systems Biology Topics Studied by Ricard, Which Are Related to the Origins of Life (Ricard, 2010a, 2010b; 2014a, 2014b)

1. Reproduction, identity, and sensitivity to signals of biochemical networks (Ricard, 2010a)
2. Networks of catalyzed chemical reactions: nonequilibrium, self-organization, and evolution (Ricard, 2010b)
3. A tentative physical model for the emergence of information and the origins of life (Ricard, 2014a)
4. Biological systems, identity, organization, and communication (Ricard, 2014b, pp. 15–34)
5. Nonequilibrium dynamics, biological systems, and time-arrow (Ricard, 2014b, pp. 184–195)

---

10 Application of Systems Analysis to Additional Complex Astrobiology Problems

We first address the long-standing contentious problem of whether viruses are alive or not by the systems approach, in which we consider virus as a part of the virus-host system.

In Chapter 11, the second complex problem we address is a specific case of panspermia, which involves the putative transfer of life between Mars and Earth. The systems analysis in this case is based on a system that includes both planets.

10.1 CONSIDERATION OF THE VIRUS-HOST SYSTEM. ARE VIRUSES ALIVE OR NOT?

10.1.1 INTRODUCTION AND OBJECTIVES

In this section, the long-standing issue of whether viruses are alive or not is addressed by utilizing the systems approach. Specifically, we consider the virus as a part of a virus-host system. Various attributes of life are readily applied to such a system. For example, while a virus particle when outside the host cannot reproduce on its own, when within the virus-host system it can do so by hijacking the host's metabolic system. Thus, the virus definitely benefits from the host in such a union. However, the host subsystem will be also evaluated. The complexity of the virus-host system will be shown, such as the existence of a win-lose scenario (the virus exploits the host to reproduce itself but kills the host in the process), a lose-win scenario (the host kills the virus before the virus could reproduce and be released), and a win-win case (the virus exploits the host for reproduction, but it enriches the host's genetic system by incorporating pieces of its genetic system). This is an important case of the systems analysis since it shows that the parts of the system may also be antagonistic.

However, the presented picture is simplistic and incomplete. The virus-host system cannot be adequately described by a single snapshot in time. The system is dynamic. To obtain a correct picture, one needs a series of snapshots of a virus and virus-host system. Further, one needs to consider an evolutionary system, which would be snapshots over a longer period of time, which involves many generations of the virus and its host.

Further, as the systems approach in general recommends, if we do not understand a system, we should investigate a broader system. While this appears counterintuitive,

DOI: 10.1201/9781003225874-10

it works! Thus, we look at the parasites other than the viruses to see if some of their features/strategies are in common with viruses.

In our presentation we move from simple to more complex, from what is a common knowledge about viruses to the more specific and specialized knowledge.

10.1.2   A Brief Background on Viruses

A virus is an infective agent whose main components are its genetic material, consisting of nucleic acid molecules (DNA or RNA), within a protein coat, called the capsid, and in some cases an outside envelope consisting of lipids (Virus: Wikipedia). Figure 10.1 shows configurations of some common viruses.

Viruses multiply only within the living cells of a host where they hijack the host's metabolic capability to reproduce. Prior to infecting a cell, viruses exist as particles termed virions (e.g., Virus: Wikipedia; Solé and Elena, 2019). Figure 10.2 shows a life cycle for viruses, which includes reproduction in the host.

The controversy about the status of viruses as being alive or not is illustrated in the following statements:

1. "… viruses today are thought of as being in a gray area between living and nonliving: they cannot replicate on their own but can do so in truly living cells and can also affect the behavior of their hosts profoundly" "…viruses,

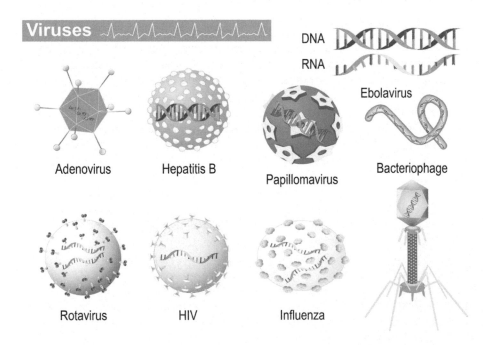

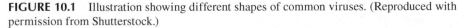

FIGURE 10.1   Illustration showing different shapes of common viruses. (Reproduced with permission from Shutterstock.)

LIFE CYCLE OF VIRUSES

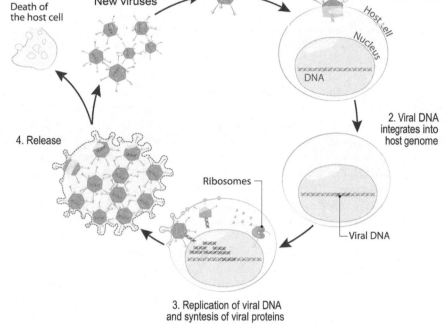

FIGURE 10.2   Life cycle of viruses. (With permission from Shutterstock.)

though not fully alive, may be thought of as being more than inert matter: they verge on life" (Villarreal, 2004).

2. "As to the question asked most frequently of all, 'Are viruses living organisms?', that must be left to the questioner himself to answer" (Kenneth Smith in 1962, as cited by Brown, 2016).

3. "The question of whether viruses can be considered to be alive, or course, hinges on one's definition of life"; "While a virion is biologically inert and may be considered 'dead' in the same way that a bacterial spore or a seed is, once delivered to the appropriate environment, I believe that viruses are very much alive"; "Alive or not, viruses are doing rather well!" (Bhella, 2016).

4. "And here is the final reflection about what viruses are and whether they are alive: considering the virion particle as the true virus is equivalent to considering a grain of pollen as a redwood or an ovulum as a human. The virion would be equivalent to the germ line of the virus while the virus factory would equate to the somatic line" (Solé and Elena, 2019).

A virion, which is the form of the virus before it invades the host, would not be alive based on the definition of life which requires self-replication. However, we notice that a virion has the potential to become alive under the right circumstances, namely after it invades the host and highjacks the host's metabolism for its reproduction. Thus, we are reluctant to designate the virion as not alive. We consider a host, which can self-reproduce and is thus alive, and a virus, when it invades the host, as two components of the virus-host system.

We next take a deeper look at the nature of the virus-host system.

The common understanding of the virus-host union is that the virus is a parasite since it exploits the host's resources for its reproduction. Sometimes, the virus kills the host. Some other times, the host fights off the viral infections, and kills the virus. The virus-host relationship thus appears to be antagonistic. However, a symbiotic relationship may exist between the virus and its host, which may result in the increased fitness of both (Bondy-Denomy and Davidson, 2014). This can happen when viral genomes are integrated into the genomes of their hosts. A recent paper describes plant virus evolution under drought conditions, which results in a transition from parasitism to mutualism (González et al., 2021). In general, viruses can transfer genes between different host species. This increases genetic diversity. Based on these and many other examples from the literature (e.g., Villarreal and Witzany, 2021; Virus: Wikipedia), we may consider some virus-host interactions to have positive results for the host.

Typically, a system is understood to be more than the sum of its parts, as evident by emergent properties which are not shown by the individual parts. From the point of view of the virus part, the virus-host system endows the virus with the ability to reproduce, which the virus does not possess on its own. The host may gain genetic diversity that may increase its fitness. However, the outcome of the virus-host system is not necessarily positive for either parts, since the virus may be killed before it has a chance to hijack the host's metabolic capacity to reproduce, or the host may be killed by the virus after the virus uses it to reproduce itself.

While this discussion is overly simplified and does not do justice to the complexity of virus-host interactions, it alerts us about the existence of antagonistic parts (or subsystems) within a system, which may result in the benefit of one part (or subsystem) and the possible detriment or even the demise of the other. This approach is simplified since the virus-host system is looked upon during a snapshot of time, in a static manner. We need to look at the virus-host system over a period of time, thus in a dynamic manner, including coevolution of both of its parts (subsystems). Such an approach will enable us to account for the apparent virus-host continuous co-existence. Such a co-existence includes also other parasite-host interactions. This topic is discussed in depth notably by Koskella (2018), and also others, who addressed some other specific features of virus-host interactions (e.g., Xue and Miller-Jensen, 2012; Villarreal and Witzany, 2021).

Based mostly on the paper on parasites by Koskella (2018), we briefly summarize the nature of the parasite-host system in general, since a virus is a parasite also. Rather than focusing on various details that are outside the scope of this book, we seek the scenarios of the parasite-host system co-existence as a model for other systems we study. We present here a simplified picture to bring out these scenarios.

The host can evolve resistance to the parasite, but the parasite can evolve to overcome this resistance. After some time, the resistance of the host toward the parasite can diminish since such resistance can be expensive to the host in terms of production of the resistance machinery. Then the parasite can attack with renewed force. This perpetuates the co-existence of the host-parasite system.

Next, we examine what is to be learned from the virus-host system, in which the parts sometimes act in an antagonistic, rather than cooperative manner. Is this an exception, or a more common occurrence in other systems? If the latter is the case, such antagonism may be silenced or destroyed by the other parts of the system, or by system itself when it forms. This raises further questions. Is there a competition between the parts such that only the chosen parts can create a system, and the other parts are discarded? We believe that such a possibility has not been sufficiently explored. Systems are typically looked at as a smooth ride in which parts somehow always interact in a cooperative manner and a new quality emerges. If so, the parts would somehow know in advance what to do and what is to be emerged. This does not seem feasible. Thus, we conclude that the parts of the prospective system are not always compatible, and some sort of selection occurs. Once the system forms, it may have parts that restrict the inclusion of noncooperative parts. Further, the nonfunction of some parts may cause malfunction of a system. Investigation of robotic systems may provide some clues, albeit reductionistic as compared to anything which is alive.

In conclusion, our common understanding of the systems in which parts interact in such a way to produce the emerging properties by some sort of cooperative interactions between the parts must be broadened. We have seen in the cases of parasite-host interaction the scenarios which are the win-lose, lose-win, and win-win.

Now we consider an additional scenario, namely, the presence of the neutral parts, which superficially does not contribute to these three scenarios. We depart from the parasite-host system and consider a chemical system, which may be more general.

The neutral part would be an unreactive species, which seemingly just "sits there." An example would be a solvent. Such unreactive parts will dilute the chemical system and slow most chemical reactions. Many more possible scenarios exist. Because of this we hesitate to call unreactive parts as neutral. They may have their way to affect the system.

Also, the co-existence of antagonistic parts, namely opposites, has been a long-term philosophical problem. It is also referred to as the unity of the opposites. We briefly introduce this topic. We consider a part of the fragment 83 by Heraclitus: "God is day and night, winter and summer, war and peace, satiety and hunger..." (Heraclitus, 2001). A plausible interpretation would be that the opposites coexist and make a whole. We can see difficulties in defining day without including its opposite, night. The same is true for war and peace, and satiety and hunger. This sort of reasoning may be applied to the case of a virus and its host in the context of the virus-host system.

Philosophy can also be helpful in framing the question if viruses are alive or not. This question is based on Aristotelian logic, in which there are only two possibilities, "yes" and "no." Both answers cannot be correct. Therefore, viruses can be either alive or not alive. However, if one applies dialetheism, which, as discussed above, is the branch of philosophy in which under certain conditions both answers can be true (Priest, 2002, 2006, 2018), to the case of viruses, then the viruses can be both alive and not alive (Kolb, 2010). In Chapter 8 on the abiotic-to-biotic transition, we also applied dialetheism when we considered the nature of the transition zone to life.

11 Transport of Life between Planets

11.1 HISTORY OF THE PANSPERMIA CONCEPT

This chapter will bring new astrobiological applications of the systems approach but will also show some unique aspects of it which may not have been sufficiently addressed in general. We consider a possibility of transferring life from early Mars to the early Earth. Interplanetary transfer of life is known as panspermia (O'Leary, 2008; Cohen et al., 2000; https://en.wikipedia.org/wiki/Panspermia), and a subtype of it, which is most feasible, is termed lithopanspermia. The latter is transfer of life by its inclusion inside a rock which shields the life from harmful radiation in space (https://en.wikipedia.org/wiki/Panspermia#Lithopanspermia).

In considering the panspermia between Mars and Earth, we first need to define the system within which such a transfer could have occurred, and its spatiotemporal characteristics. The system comprises obviously these two planets, but only at a particular time span, when the habitability of Mars and Earth coincided. The transfer of life depends also on the spatial factors. If life was ejected from Mars, it would generally travel through space over a long period of time before impacting Earth, which depends on the pathway of travel which, in turn, depends upon the energetics of ejection of life from Mars. The viability of life during the transit depends on the length of time of the transfer. In addition, the space between Mars and Earth is not uniform but has its own local features which may influence the outcome of transfer of life between these two planets. For example, the space around Mars would have been the Mars' atmosphere, while Earth had its own. They were not the same. In addition, the space close to these two planets would be experiencing local gravitational and other forces, which the life-containing material would have to overcome, during the ejection from Mars, and/or landing on the Earth. More aspects of the "empty space" between Mars and Earth need to be considered, such as the existence of various smaller objects that may be periodically found in space, and which may derail, absorb, or destroy the life in transit. Further, the Mars-Earth system changed over time since its subsystems, namely Mars and Earth, took different historical paths. Life may have started on Mars, may have transferred to the Earth, and thus may have spanned both planetary subsystems at a particular period of time. As Mars became less and less habitable, and the opposite was true for Earth, these two subsystems, as far as life is concerned, diverged. They developed their own habitability history (Clark et al., 2021).

In this chapter, we elaborate on the spatiotemporal features of this system and bring more specifics and relevant references to make our case.

The panspermia concept has a long history. Anaxagoras, the Greek philosopher who lived in the 5th century BC, believed that the universe is made of an infinite

DOI: 10.1201/9781003225874-11

number of seeds (*spermata*). He introduced the term panspermia, which literally means "seeds everywhere." Although the available fragments of his work do not contain a description of the nature of the "seeds," the latter are generally interpreted by the philosophers to mean the source of life, such as in agriculture, which Anaxagoras was for sure able to witness (O'Leary, 2008; Cohen et al., 2000). Like other works by the early Greek philosophers, Anaxagoras' panspermia hypothesis came only from thinking, without any concrete experimental evidence, and thus constitutes an undeveloped hypothesis.

The panspermia hypothesis was taken up and developed further by several scientists at the end of the 19th century and the beginning of 20th century (Arrhenius, 1908; Longstaff, 2015; Tirard, 2013). For example, in 1871 Sir William Thomson (who later became Lord Kelvin), a physicist, claimed that "germs" (microorganisms) coming from space via meteorites originated life on Earth. Herman von Helmholtz, another physicist, adopted this view in 1884. In 1903, Svante Arrhenius, a chemist, proposed that the spores can be accelerated by the radiation pressure of sunlight, and thus can be transported in space and seed life on the planet on which they land.

Later, the panspermia hypothesis considered not only the interplanetary but also the interstellar transport of life via the microorganisms or spores. Landing of such "spermata" on another habitable site would spread life throughout the universe. The concept of "directed panspermia," the purposeful transfer of life by the intelligent beings, was also proposed (Crick and Orgel, 1973; Crick, 1988). At this point of history, the intelligent beings could be us.

What started as an idea and belief in the old Greece is now taken very seriously and much experimental work has been done to address it. This was prompted notably by the space missions to search for life in our Solar System. Such missions had to make sure not to spread the Earthly life to the other planets, where life was sought. Further, there was a fear of contamination of the Earth with foreign life which could be brought back by returning spaceships. This necessitated the development of a Planetary Protection program (Conley, 2019), the protocols for which are accepted (at least in principle) by all the major space-faring nations.

11.2   PLANETARY SYSTEM TRANSFERS

A central experimental approach has been to study the survivability of microorganisms and their spores in space. These can be accomplished at the International Space Station. The microbial spores were placed on the outside of the ISS, where they were exposed to vacuum desiccation, low temperature, and harmful radiation, which are all conditions not friendly to life. However, the spores showed resilience and survivability under these conditions (Horneck et al., 2016), which implied that they could survive travel in space for a limited time. Further, when the spores were encased in rocks, which shield them from radiation, they were able to survive much longer, which increased the likelihood of a successful interplanetary transport of life. This type of panspermia is termed "lithopanspermia" (from Greek "lithos" which means stone).

Additional support for the feasibility of panspermia is provided also by the various examples of the transport of geologic material between the planets by exchange of meteorites (Treiman et al., 2000). As one example, several hundred meteorites from Mars have been found on the Earth. If the rocks that are exchanged between the planets contain bacterial spores, lithopanspermia becomes a possibility (Horneck and Bäcker, 1986; Clark, 1986, 2001, 2002; Melosh, 1984, 1988; Mileikowsky et al., 2000).

Thus, a combination of the survivability of spores inside the rocks during the space travel, and the exchange of geological materials between the planets, makes panspermia feasible.

However, this feasibility needs to be evaluated considering various adverse factors. As just one example, the galactic cosmic rays (GCR) will penetrate several meters into rock and can sterilize the spores during a prolonged exposure. In the next section we present a detailed evaluation of lithopanspermia between Mars and Earth's planetary system.

11.2.1 Possible Panspermia between Mars and Earth

As an exemplar of a more difficult-to-relate approach of our generalized systems block diagram, as representation of a system, we examine Mars-Earth lithopanspermia in some detail. Each of the blocks in the diagram represents a grouping of similar topics which should be identified and evaluated for its level of criticality and participation in the overall "success" or "function" of the system.

Let us begin with "Energy." Accomplishing the transfer of rocky material requires that it first be excavated from the martian surface and then sufficiently accelerated to escape the gravity of Mars (5 km/s escape velocity). Although there are no known forces on Mars that can accomplish this, an impact by a small asteroid, comet, or other similar interplanetary object could do it. Such objects impact Mars at velocities of tens of km/s which are sufficient to break bedrock and eject the fragments to large distances during formation of the impact crater. Mars has more than one hundred thousand impact craters large enough to have ejected particles into trans-Earth space (Robbins and Hynek, 2012). More impacts have occurred because some craters are now covered by sediments. Similarly, the Moon has obviously also been subjected to such bombardments. In fact, a number of meteorites have been identified as having come from the Moon, now that we know from the Apollo and Luna missions what the compositions of lunar materials are. Also, the Earth has been subjected to extensive bombardments, although only ~100 craters can be seen today due to obscuration of most original craters by sediments, lava, water, etc.

Initial suggestions that a few meteorites of unusual composition in the vast collection must have come from a planet, and possibly from Mars, were greeted by abject skepticism by the geophysicists because it was thought not possible to eject a rock from the martian gravity well without melting the minerals within it. Then, the Melosh mechanism (Melosh, 1984, 1988) was proposed, which takes advantage of interferences in shock waves to result in a fraction of ejected material to be launched by the mechanism of spallation, resulting in igneous or other strong rocks being accelerated to high velocities without totally crushing or melting. Weaker materials, such as sediments, weathered rocks, sand, or soil will simply disintegrate and not be efficiently launched. The crater-forming impacts capable of ejecting potential martian meteorites for hitting Earth are enormously energetic, exceeding 50 Megatons of equivalent explosive energy, corresponding to the most powerful explosion by thermonuclear fusion (hydrogen bomb) ever tested.

Thus, the transfer of microorganisms from Mars to Earth must be preceded by this most energetic event. However, because the spalled rocks that are ejected intact leave the planet within seconds, they are heated very little and can escape with a precious "payload" of organisms. These microbes must be in a quiescent state or very quickly transform to a form that can withstand dehydration by exposure to the very high vacuum of space. Many bacteria are capable of forming spores that require little or no nutrition and no significant energy source in order to survive for thousands, and perhaps millions of years (Wörmer et al., 2019, and references therein), only to become reactivated in only a matter of hours or days once exposed to water and nutrients. Such spores are ideal payloads for transport between the planets and planetary bodies (e.g., large moons of the giant planets) because they tolerate high vacuum and generally are more radiation resistant.

However, the travel time between the planets is on the order of a few Myr (Million years) – only about 5% reach Earth in less than 10 Myr in space. In comparison, missions to Mars launched on rockets from Earth are highly optimized for launch date and use precisely aimed launch vehicles (rockets) and trajectories to accomplish the trip in less than 1 year. However, these special "Hohmann transfer ellipse" trajectories would be extremely rare for naturally ejected material from Mars. Travel time is quite important, because although bacterial spores can well withstand the cold temperatures and vacuum of space, they will be gradually killed by the ionizing radiation from galactic cosmic rays and solar flare radiation events. Being inside a large rock will be helpful, but an analysis of known martian meteorites, all of which have been in space for 1–15 million years, showed that the radiation doses would be several to hundreds of megarads (1 megarad = 10 kGy), enough to completely sterilize even a huge colony of typical bacteria (Clark, 2002). Martian meteorites found to date have been much, much less than 1 m in size, down to a centimeter, with the smaller meteorites much more numerous than the larger ones.

Once reaching the Earth, the organisms must survive the heating during passage through the atmosphere, and the impact of hitting the ground. Very large meteorites will not be effectively slowed by the atmosphere, and will impact at multiple km/s velocities, which will generally sterilize the sample by the shock and heating of impact. The known martian meteorites are smaller and consist of strong rock. Although each produces an impressive "fireball" during their high-speed entry, they actually are heated on only a thin surface layer and remain cold inside.

Meteorites may have broken apart in pieces from the stresses of entry or upon hitting the ground, but organisms surviving the space environment will generally survive these as well. Such organisms then must, however, adapt to the new environment. As spores, they must germinate into their active form and be able to subsist in terms of obtaining nutrients and energy from their surroundings. This will be difficult for bacteria deep within the rock, perhaps sealed in by the outer layer of the rock which melted during entry.

For panspermia, the sensors are in the bacteria, i.e., the payload. Spores are inactive in most respects, but they still can detect favorable conditions, such as the return of moisture or water and nutrients. Without some sensing capability to trigger germination, to exit the spore state, the organisms would not return to their prime function, i.e., reproduction.

Bioloads of igneous rocks are generally very low, compared to sediments. Yet, it is the strong, competent rocks that can withstand the loads during launch. A low population of organisms most likely also means less diversity, which may be quite important if most organisms from Mars do not happen to be attuned sufficiently to their new home on Earth.

Many bacteria are capable of self-locomotion, which provides the motility to escape the original meteorite that transported them to Earth, enabling them to populate the new external environment, making the first step for creating a planetary biosphere. Thus, this overall system has both mega- and micro-transport systems which are critical to overall success.

Now, let us systematically go through the subsystem list of functions to see how it affects the probability of the overall system function of transferring life from one planet to another.

11.2.2 POWER AND ENERGY

With respect to the energy necessary to escape Mars, the existence of interplanetary debris is essential to provide the impactors. In general, these objects will have high velocities relative to Mars, and the martian gravity field will accelerate them further.

The gravitational attraction by Earth will generally add energy to the impacts, which is not a positive factor. However, the Earth's atmosphere will "bleed off" kinetic energy by friction heating, and thereby soften the blow of landing.

From the microbe's standpoint, they will presumably have had adequate sources of multiple energy on Mars (e.g., chemical or solar energy sources). And if they land in a similar environment on Earth, they will have the best chance to succeed.

11.2.3 CONTROL AND DATA

The microbes themselves have adequate control of their capabilities. And the DNA in their genome is the repository of the "data" that tells them how to operate and provides the plans and commands for reproduction that will enable future generations.

11.2.4 SENSORS

As mentioned above, the microbe must be capable of sensing favorable vs unfavorable environmental conditions to know when to transition between vegetative and endospore states.

In order for quorum sensing to be an advantage, each microbe must be capable of sensing numerous signals. In addition, if the organisms have the capability to self-locomote, then sensory capabilities would allow photo- and chemotaxis, or other responses that may be advantageous or even essential to finding a local niche on the new planet where they could prosper.

11.2.5   STRUCTURES AND MECHANISMS

The structures that must be involved to enable a successful lithopanspermia are at both the macro- and micro-scale. Strong, igneous rocks can certainly be transported, as demonstrated by the more than one hundred known meteorites from Mars. However, we have found no weak materials from Mars, such as sediments, cemented soils, and salts.

The organisms must survive not only the shock and acceleration forces of launch but also the decelerations during entry at Earth as well as the impact onto the ground. Fortunately, microorganisms can be extremely robust against mechanical insults. And endospores can be even more hardy because of their solid interiors sufficient to withstand high shock levels (Burchell et al., 2004).

11.2.6   IMPORT-EXPORT-STORAGE

The Import/export function of the block diagram is also critically tied to the landing environment. The microbial complement must have access to key nutrients, such as those containing the CHNOPS and other key elements (e.g., Fe, Mg, K). It goes beyond stating that an adequate supply of H_2O is essential. Depending on the metabolic needs of the microbial community, sunlight may be required, and a host of special molecules may be necessary, such as CO_2, from the atmosphere, and reduced forms of nitrogen in soil or water.

11.2.7   COMMUNICATIONS

Transfer of information between members of a bacterial consortium can be valuable in the transition to survive a new environment. For example, quorum sensing within a community can be a valuable method of adapting to a new, alien environment.

11.2.8   INTERNAL ENVIRONMENT

Maintaining an internal environment conducive to survival, and especially to growth and reproduction, is far easier if the external environment is similar to the one they are already evolutionarily adapted to. Thus, the external environments on Mars need to be similar to the ones at wherever the landing site is on Earth. This requires that the same favorable conditions exist on both planets. On the other hand, the geology of early Mars can indeed be quite similar to that on early Earth, given the similarity in surface minerals (basaltic) and the probability of comparable early atmospheric compositions.

The thermal component of most systems is of significance here also. We have already shown that the thermal environment is survivable for the ejection, the entry at Earth, and the thousands or millions of years of traverse in space. However, even the hardiest of microorganisms have a limited range of temperature over which they are highly functional and reproductive. From psychrophiles which thrive at temperatures up to 15°C and down to the freezing point of water, to hyperthermophiles which survive, grow, and reproduce at temperatures above 80°C and higher, there are no bacteria that thrive over the entire temperature range of liquid water, from freezing to boiling (e.g., Oren, 2019). Thus, if martian psychrophilic organisms are delivered to hot springs or a tropic climate, they may stall or die before they can take hold as a population. Furthermore, the boiling point of liquid water depends critically on pressure. For example, at a martian pressure of only 15 mbar, the boiling point is +13°C, which means martian organisms would be psychrophiles and would need to land in a cold region on Earth to prosper.

11.2.9 MOBILITY

The transfer between Mars and Earth could be thought of as transportation which is not under control of the organisms themselves. Rather, they are passive passengers. However, once they landed on Earth, motility of the microbes would be an advantageous function for survival in a new environment. It can help them exit their transport system (the rock), and enable taxis responses in search of food or sunlight.

11.2.10 NAVIGATION

A lithopanspermia subsystem would have no means to ascertain its precise location, nor to target a specific planet. If it did have navigational knowledge, and if it did have a propulsion system, then the trip could be shortened. This would be the case for artificially directed panspermia, but is not the case for natural panspermia.

11.2.11 REPRODUCTION

It is unlikely that the bacterial passengers would be able to reproduce during the transit, because the interior of the rock would be very cold, and there also would not be a source of water in the liquid state. Once landed on Earth, it would be essential that the hitchhikers be able to reproduce and evolve any adaptations needed to guarantee long-term survival. Ideally, the meteorite should touch down in water (lake, ocean) or in wet soil. And the organisms would need to be able to take advantage of whatever nutrients were available in soil, water, or atmosphere. However, although it is the most likely landing location, Earth's enormous ocean is relatively poor in nutrients.

11.2.12 DEFENSE

The most important threat to the survival of organisms inside rocky material during its time in deep space before serendipitously encountering Earth is inactivation by ionizing radiation. Although it is highly hazardous, solar UV is easily shielded so that organisms below the surface by less than 1 mm will remain viable. And assuming the organisms can become sufficiently transitioned to extreme dormancy such that they do not need food, water, or warmth, there is still the issue of death by GCR (Galactic Cosmic Radiation) or SPE (Solar Particle Event) radiation, or both.

As noted above, many bacterial species have significant resistance to ionizing radiation, up to a level of ~1 Megarad. Some unique microorganisms, such as

Deinococcus radiodurans, are even more resistant because they can repair radiation damage as it occurs. However, this organism is an obligate aerobe, and Earth's atmosphere would have only extremely low concentrations of O_2 before a biosphere was formed. A few other organisms are now known to have extreme radiation resistance, but they have other special needs, such as phototrophy.

The radiation dose that lithopanspermia organisms would experience in space depends on how deep they are in the ejected rock they inhabit, to avoid sterilization by ≥ 1 Mrad exposure to GCR and SPE. Detailed calculations show that a typical bacterium would need to be buried at the center of at least a 10 cm diameter rock and transit from Mars to Earth in less than 10,000 years (Clark, 2002). Trajectory calculations show that only one rock in every 1,000 that are ejected from Mars will reach Earth in that length of time. For times in space of more than 1 million years, even if the organisms survived vacuum desiccation, they would still need to be at the center of a boulder at least several meters in diameter. Large meteorites are much rarer than smaller ones, by about four orders of magnitudes for the abundant ones of a few centimeters diameter compared to multi-meter boulders such as the rare case of the Chelyabinsk meteor nearly 20 m in diameter (which was not from Mars).

Even though the formation of an impact crater on Mars has the potential to launch millions of small particles, many of these will be re-captured by Mars or suffer some other fate before an opportunity to actually hit the Earth. Organisms deep inside the rock would need to survive entry and landing, but these insults are minor. Entry heating of a meteorite can be so severe that it melts the outer layer to produce a dark "fusion crust." However, the interior of the rock remains cold. On the other hand, this fusion crust can seal over the normal fissures formed in a rock by shock of the launch and make it difficult for the transported bacteria to escape from the rock to become active on Earth.

11.2.13   INTERCONNECTS

Interestingly, the connectivity of stages in the pathway of transfer from one planet to another is essentially independent. Higher speeds of initial impact will eject more material and therefore may place more material onto Earth-crossing trajectories. That could, then, increase the approach velocity to Earth and result in a somewhat lower probability of successful entry and landing.

11.2.14   BOUNDARY

The boundary for a lithopanspermia function is our Solar System. Although the key portions are the surfaces of Earth and Mars, the trajectories for the ejected particles can span not only the space in-between the two orbits, but their aphelia can be further from the Sun than Mars orbit, and some perihelia will come closer to the Sun than Earth's orbit (with the possibility of overheating).

However, this is a case where boundaries change. Initially, the traveling organism is restricted to the surface of Mars. And once it lands on Earth, it is likewise bounded by the vicinity of the local environment, until its descendants are spread by wind or water currents.

11.3   MARS-TO-EARTH LITHOPANSPERMIA

In the opposite case of Earth-impact and then transport of microbes to Mars, there are similar functions and advantages and disadvantages. However, the probabilistic numbers are different.

In general, it appears more likely that Mars organisms could reach Earth, than the opposite. Mars is a smaller target, which could mean less impacts. However, Mars is closer to the asteroid belt, which is generally a greater source of impactors, although in the early Solar System, potential impactors may have been more evenly distributed.

It is significantly more difficult for surface materials to escape Earth, because of our higher gravitational strength that must be overcome (5.0 km/s to escape Mars; 11.2 km/s to escape out of the gravitational well of Earth). In addition, Earth's much denser atmosphere is a greater impediment to successful escape from the gravitational influence of Earth.

The Earth may have actually been a less-favorable surface environment than Mars for an early origin of life (Benner and Kim, 2015; Clark et al., 2021). Earth may have been primarily a water-world, with little land and a nutrient-poor ocean. Interestingly, if martian life is ultimately to be discovered, and if it can be shown to have tell-tale similarities to terrestrial life at the biochemical scale, then the next question will be "on which planet life started?"

11.4   INTERSTELLAR PANSPERMIA

The idea of panspermia between two different solar systems is far, far less likely. For example, our nearest star, Proxima Centauri, is nearly 5 light-years in distance from Earth, compared to Mars varying from only 3 to 22 light-minutes from us. In other words, in straight-line distance the nearest star is half a million times farther away! Natural panspermia seems untenable. But directed panspermia using our advanced technologies could be considered.

Suppose, however, we could manage to escape a spaceship from our Solar System with a net velocity of 100 km/sec (223,000 mph). It would still take 125 centuries to reach Proxima Centauri. But worse would be the fact that it would be necessary to slow down in order to rendezvous with some planetary target.

A very large mass of spacecraft would be needed, to provide an enormous amount of shielding to protect the cells from GCR radiation over the very long term. Perhaps with technologies for keeping the cells actively repairing damage and producing new organisms, this danger could be overcome.

However, at these extremely high speeds, if the spaceship encountered even a small particle, it could suffer extreme damage due to the collision at these very high speeds. In the military, this is termed the "kinetic kill" mechanism. Countermeasures are possible, using the Whipple shield concept of multiple layers spaced apart to successively break the particle into smaller particles, each of which spread out and progressively lose energy as they create holes in successive shields. Such shields were designed and used to protect the spacecraft for the Stardust sample return mission to comet Wild 2 (Sandford et al., 2021).

Studies are indeed underway to see if it is possible to construct a system that could make an interstellar journey in a reasonable time. One concept is to use powerful Earth-based laser beams shined on very lightweight spacecraft with receptor target sails, the so-called "Breakthrough Starshot" project (Parkin, 2018).

Apart from the intelligent design of a special spaceship, it does not at all seem feasible for interstellar lithopanspermia to have even a tiny chance of succeeding. There would need to be a natural mechanism for accelerating the transporting rock to these extremely high speeds. Yet the rock must be a huge boulder to shield the organisms from GCR and collisions with interstellar matter. And there would be no natural way to slow down the vehicle once it reached its target, resulting in a "burn-up" end to the journey.

11.4.1 INTERSTELLAR COMMUNICATIONS OF INTELLIGENCE

If, however, it remains impractical to transport matter, especially life, across interstellar distances, there is something else we could transport. We can transport information, i.e., communicate. As part of the Search for Extra-Terrestrial Intelligence (SETI) by observing signals coming from outside our Solar System (e.g., Korpela, 2019), we can not only seek to detect communications from life in other stellar systems, but we could share our knowledge of the chemistry and physical makeup of a life system by sending signals, i.e., METI (Messaging Extra-Terrestrial Intelligence) (https://en.wikipedia.org/wiki/METI" https://en.wikipedia.org/wiki/METI) containing data about our own life forms.

For example, we already know from astrophysics studies of spectral emissions in deep space that the set of elements and their isotope stabilities is essentially invariant across the universe. We could transmit our knowledge of that data set as binary signals, working our way up the Periodic Table. For example, we could transmit the binary sequence for the number of protons and neutrons for each stable isotope between, say, hydrogen and iron. From there, it would be possible to transmit the short-forms of molecular compositions in terms of their atoms. Obviously, we would want to send the compositions of key amino acids for life and the nucleotides that make up our DNA and RNA. More complicated, but possible, would be to communicate the stereochemistry of each key compound. Then, it could be attempted to send the compositions of ATP and some Fe-S proteins, as well as the structure of DNA with an example of a stretch of base-pair sequences, perhaps the coding of one of the key sequences.

Whether it would ever be practical to send enough information for a target civilization to actually construct a viable primitive bacterium in their laboratories from a transmitted set of instructions is questionable – especially considering that we cannot yet do that even ourselves. But it would be highly valuable to identify our principal biochemical components. Hopefully, the other civilization also thought about that strategy and has already transmitted a bit stream that has the potential to be decoded accordingly. This could immediately assuage our intense interest in not only whether there is there life elsewhere, but if so, what is it made of and how similar is it to us?

12 Retrospective on the General Applicability of Our System Functional Block Diagram

With this book, we have endeavored to bridge systems engineering principles and its approaches for analyses over to the disciplines of astrobiology, with emphasis on the grand questions of the origin and evolution of life. While this is not meant to be an exhaustive survey of the topics, it is hoped that these exemplars evoke a wide range of subsystems issues and benefit the insights into astrobiological topics.

Our focus in this retrospective is to summarize the general applicability of the elements of our system functional block diagram, as shown in Figure 4.6. The reader may have noticed the repeated application of each of the subsystems of this diagram to seemingly unrelated complex problems, such as a planetary rover (Section 4.4), the physiology of a human being (Section 4.5), the macrobiont (Section 7.6), and lithopanspermia (Section 11.2).

Although most studies are focused on much narrower issues, we believe it is useful for the specialists to ask themselves how their area of specialty fits into the overall system at large. In Table 12.1, we provide some of the questions which arise, while previous chapters have provided some qualitative answers to these questions which can lead to more quantitative and consequential studies.

Becoming mindful of the overall system's functions having dependencies on the various sub-functional areas can help identify where there may be other relevant factors not previously appreciated. To the extent that a subfunction can be computationally modeled, it becomes possible to better understand the influence of that subfunction, as well as to identify the combined influence of combinations of key subfunctions.

Systems engineers and their teams also bore down within each subsystem to determine its performance limits, especially under the actions of plausible stressors. Thus, not only do the engineers test each and every engineered box for its limitations to thermal stress, but they also place the overall system into test chambers that simulate temperature, vacuum, and operational extremes (the so-called "T-Vac tests").

In chemical and biological systems, much can be learned by varying several conditions, including temperature, pH, Eh, ionic strength, and concentrations of specific favorable as well as unfavorable chemical constituents. Although this could lead to exhaustive testing of questionable value, there can instead be pathfinder tests with less granularity in variables and their combinations. Once significant impacts are

DOI: 10.1201/9781003225874-12

TABLE 12.1

Subsystems and Their Functions

Subsystem	Some Key Functions
Energy and power sources	What energy sources available? (chemical, UV/VIS, thermal, gravitational, electrical, mechanical?) How stored? How distributed (power)?
Control and data	How are subsystems controlled? Feedback loops? Hysteresis? Tipping pts? Nonlinearities? What information ("data") is stored and how is it used?
Sensors	What sensors are available? How are they converted (transduced) into signals to other subsystems?
Structures and mechanisms	What is the mechanical framework of the system? What mechanical movement devices (mechanisms) are involved?
Import-export-storage	What physical masses are moved or stored (nutrients, wastes)? Why and where are they stored, and how are they transported (in or out)?
Communications	How is information transferred within the system and interchanged with the external world? To what extent is the system under external control?
Internal environment	How is the environment inside the system different from the outside environment, and how is it maintained (homeostasis)?
Mobility	Is the system able to translocate, either under its own mechanical power or transported by external forces?
Navigation	Can the system determine its location in space, and can it follow a desired path if it has the ability to guide itself or self-locomote?
Reproduction	Is the system self-reproductive? If so, how? Or, does it depend on external actions to be reproduced?
Defense	What compensating activities are intrinsic to the system that favor self-preservation? What are their components (mechanical? chemical?)
Interconnects	How are various subsystems connected with one-another? Are they only under a master controller, or some lower-level connectivities?
Boundary	What is the nature of the boundary that sets the system apart from its environment (rigid versus semipermeable)?

found, there can be more detailed but constrained testing for the key variables identified, or the results can be used to point to future work that may be taken up when more resources are available, or outside investigators become interested.

In reality, subsystems generally also have their own subsystems. There may be sub-sub-sub-subsystem levels, such as a Si-based semiconductor chip that is sealed inside a piece-part that is on a circuit board, connected to a motherboard, that is inside a box. Or, consider that there are not only ribosomal-, messenger-, and transfer-RNAs that are used to convert information in DNA into protein amino acid sequences, but there are also other forms, such as micro-RNAs that exert control functions, as well as catalytic RNAs (ribozymes), and even noncoding RNAs.

Also, especially in the biological world, there are entities that are grossly multi-functional. This can create unavoidable, complex feedback loops which are particularly difficult to analyze without determining functional dependencies and using advanced mathematical modeling and/or machine learning to tease out overall responses to complex stimuli or inputs.

When a new artificial system is developed from scratch, such as a race car, airplane, or planetary rover, the engineers automatically adopt an electronics-based logic system, operated with electrical power. Advanced high-energy-density batteries invoke a somewhat complex chemical system, but that chemistry is relegated to the confines of the battery boundary. Otherwise, both the box interiors and their interconnections are based on the flow of electrons to convey information, power, analysis, and even mechanical movement. Even the sensors chosen transduce their detections into electronic outputs.

However, by systems analysis, it is possible to treat a metabolic subsystem or the brain (or its sub-portions) as an individual subsystem that has overall inputs and outputs that reflect its main functions. Even this can be challenging, of course, simply because the capabilities of brains and metabolic pathways evolved in a pseudo-random path from lower-level functionality to a higher level of complexity, sophistication, and interdependence on other sub-subsystems.

Nonetheless, the systems approach does just that, even for the most complex airplane or spacecraft. Although it may be tempting to create an overall model that incorporates every single individual component inside the system, this simply complicates and bogs down the model when trying to understand how it works. For a modern passenger airplane, that would involve thousands of "parts." Likewise for a spacecraft, modeling the complexity down to the tiniest innards, such as electron flow inside a semiconductor, and combining such models into one grand scheme of calculation would be overpowering and, in the end, unnecessary.

An analogy is when, early in the space program, there was initially an attempt to calculate the probability that an overall system would fail by combining the probabilities for all the individual parts that they would perform outside the limits of the manufacturer's specifications for each of the smallest parts. Invariably, such calculations predicted that success would be much more unlikely than seemed reasonable. With time, the practice of determining the probabilities of failure and success was changed. Instead of analysis at the smallest part level, it was decided to test units at the box assembly level. Such tests were conducted for long periods of time on several identical units to see how many failures occurred. So-called "accelerated" tests to failure were also conducted by stressing the test units – for example, by operating them at higher temperatures than they would experience in the real system. This led to the characterization of units in terms of MTBF (Mean Time Between Failure). It was also learned that this number could be quite high, with values in the thousands of hours, but only provided that the typical early-onset failures were neglected. These so-called "infant mortality" cases were when there were apparently initial faults in the manufacture of those units. This was analogous to cases in biology where those newborn organisms which were apparently "normal" would survive much longer than the newborn with one or more initial deficiencies. To tease out the possibility of infant mortality, spacecraft are typically operated for a few hundred hours on the ground, before being shipped to the launch site.

Some lesson learned from this experience is that when studying any complex system, even a small defect can eventually have serious consequences. And, systems with anomalies may not be fruitful for study of the overall system, with the exception that the anomaly may bring to light some dependencies of other subsystems as well as the overall system on that particular subfunction.

Another lesson is that small perturbations resulting in small changes in performance may not be sufficient to determine overall capabilities and expectations for the system as a whole, because when one subsystem falters, other subsystems may be called upon to compensate and may, one way or another, be relatively successful.

Altogether, a system view from the highest level of performance has the potential to discern the deconvolution of the underlying subfunctions which comprise it and enable its robustness. Research at a particular subsystem level can benefit from discovering or hypothesizing the functions needed for that subsystem as context for revealing the capabilities and limitations of that subsystem. With enough diverse efforts probing the various subsystems, a broad picture and ultimately a detailed model of that system will become available. And with it comes a better characterization of what each of the subsystems is capable of, and how it contributes to the overall function of the next higher-level subsystem or system itself.

With the advent of machine learning algorithms, it is becoming possible to develop a complex, parametric mathematical model of a subsystem simply by operating it under a variety of conditions, with carefully documented inputs and outputs to document the data to be analyzed. Such models can then be used to predict subsystem behavior in off-nominal circumstances, although these then need to be verified for extreme cases if the artificial model does not capture the actual underlying interdependencies of the variables involved.

Independent of the method of analysis, it is important to determine the nonlinearities within the subsystem performances. Positive and negative feedback processes, replete in metabolic systems, are especially crucial to identify and evaluate because their influence and complexity can be paramount. Likewise, control is partially active, through the judicious use of catalysts to overcome sluggish reactions and sources of chemical energy such as the ATP-to-ADP transition. Control can also accrue passively when reactants are available with thermodynamically favorable chemical pathways having low activation energies.

With the passage of time, the system may change (and ultimately "die"). To what extent is it susceptible to "wear-out" versus spontaneous isolated failures within the system? And must the system ultimately experience some highly unfavorable environment(s) that could be catastrophic to survival?

Just as mathematics is more than following rules but is also especially a way of thinking that takes practice and insight, so it is with systems analysis. We hope these discussions have been thought-provoking, useful, and motivating. Although it may be true that the human mind typically focuses on the immediate minutiae instead of the holistic patterns, this does not mean that we cannot appreciate both "the forest and the trees" at the same time. Our point is that it is not enough to study everything about the tree without realizing it is actually part of a larger, complex forest ecosystem, which in turn is embedded within a global biosphere, and which exists because of the favorable geosphere and atmosphere that surround it. Life exists within a thin veneer just a few kilometers high and a few kilometers below ground level on a planet that is thousands of kilometers in diameter of sterile mineral matter. Nonetheless, in spite of this apparent fragility, the biosphere overall has survived over geological time because it is a versatile, dynamic, and adaptive megasystem that, despite its great internal complexity but enormous redundancy and diversity, is unlikely to succumb to all but the most dire of circumstances.

References

Abramov, O; Bebell, KL; Mojzsis, SJ. Emergent bioanalogous properties of blockchain-based distributed systems. *Origins of Life and Evolution of Biospheres* **2021**, 1–35. https://doi.org/10.1007/s11084-021-09608-1.

Alon, U. *An Introduction to Systems Biology, Design Principles of Biological Circuits, 2nd Ed.*, 1st ed. 2006, CRC Press: Boca Raton, 2020.

Arnold, RD; Wade, JP. A definition of systems thinking: A systems approach. *Procedia Computer Science* **2015**, *44*, 669–678. https://doi.org/10.1016/j.procs.2015.03.050.

Arrhenius, S. *Worlds in the Making: The Evolution of the Universe*, Harper & Row: New York, **1908**.

Ashkenasy, G; Hermans, TM; Otto, S; Taylor, AF. Systems chemistry. *Chemical Society Reviews* **2017**, *46*(9), 2543–2554. https://doi.org/10.1039/C7CS00117G.

Ashwell, K. *Anatomy & Physiology*, Barron's: New York, **2016**.

Autocatalysis. https://en.wikipedia.org/wiki/Autocatalysis.

Autocatalytic set. https://en.wikipedia.org/wiki/Autocatalytic_set.

Baross, JA. The rocky road to biomolecules. *Nature* **2018**, *564*, 42–43. https://doi.org/10.1038/d41586-018-07262-8.

Baross, JA; Anderson, RE; Stüeken, EE. The environmental roots of the origin of life. In *Planetary Astrobiology*, Meadows, VS; Arney, GN; Schmidt, BE; Des Marais, DJ, Eds., University of Arizona: Tucson, pp. 71–92, **2020**.

Bartlett, S; Wong, ML. Defining lyfe in the universe: From three privileged functions to four pillars. *Life* **2020**, *10*(4), 42. https://doi.org/10.3390/life10040042.

Bar-Yam, Y. General features of complex systems. *Encyclopedia of Life Support Systems* **2002**.

Becker, S; Feldmann, J; Wiedemann, S; Okamura, H; Schneider, C; Iwan, K; Crisp, A; Rossa, M; Amatov, T; Carell, T. Unified prebiotically plausible synthesis of pyrimidine and purine RNA ribonucleotides. *Science* **2019**, *366*(6461), 76–82. https://doi.org/10.1126/science.aax2747.

Becker, S; Schneider, C; Crisp, A; Carell, T. Non-canonical nucleosides and chemistry of the emergence of life. *Nature Communications* **2018b**, *9*(1), 1–4. https://doi.org/10.1038/s41467-018-07222-w.

Becker, S; Schneider, C; Okamura, H; Crisp, A; Amatov, T; Dejmek, M; Carell, T. Wet-dry cycles enable the parallel origin of canonical and non-canonical nucleosides by continuous synthesis. *Nature Communications* **2018a**, *9*, 163. https://doi.org/10.1038/s41467-017-02639-1.

Benner, SA; Kim, HJ. The case for a martian origin for Earth life. In *Proceedings of the Instruments, Methods, and Missions for Astrobiology XVII*, International Society for Optics and Photonics: Bellingham, WA, vol. 9606, p. 96060C, 28 September **2015**.

Bettelheim, FA; Brown, WH; Campbell, MK; Farrell, SO. *Introduction to General, Organic, and Biochemistry*, 9th ed., Brooks/Cole, Centage Learning: Belmont, CA, **2010**.

Bhella, D. Yes, viruses are alive. *Microbiology Today*, *43*(2), 58–61, 10 May **2016**. https://www.microbiologysociety.org/publication/past-issues.

Bondy-Denomy, J; Davidson, AR. When a virus is not a parasite: The beneficial effects of prophages on bacterial fitness. *Journal of Microbiology* **2014**, *52*(3), 235–242. https://doi.org/10.1007/s12275-014-4083-3.

Bonfio, C; Russell, DA; Green, NJ; Mariani, A; Sutherland, JD. Activation chemistry drives the emergence of functionalised protocells. *Chemical Science* **2020**, *11*(39), 10688–10697. https://doi.org/10.1039/D0SC04506C.

Britvin, SN; Murashko, MN; Vapnik, Y; Vlasenko, NS; Krzhizhanovskaya, MG; Vereshchagin, OS; Bocharov, VN; Lozhkin, MS. Cyclophosphates, a new class of native phosphorus compounds, and some insights into prebiotic phosphorylation on early Earth. *Geology* **2021**, *49*(4), 382–386. https://doi.org/10.1130/G48203.1.

Brown, N. No, viruses are not alive. *Microbiology Today 43*(2), 58–61, 10 May **2016**. https://www.microbiologysociety.org/publication/past-issues.

Brunk, CF; Marshall, CR. 'Whole organism', systems biology, and top-down criteria for evaluating scenarios for the origin of life. *Life* **2021**, *11*(7), 690. https://doi.org/10.3390/life11070690.

Burchell, MJ; Mann, JR; Bunch, AW. Survival of bacteria and spores under extreme shock pressures. *Monthly Notices of the Royal Astronomical Society* **2004**, *352*(4), 1273–1278. https://doi.org/10.1111/j.1365–2966.2004.08015.x.

Capra, F. *The Web of Life*, Anchor Books, Doubleday: New York, **1996**.

Capra, F; Luisi, PL. *The Systems View of Life, A Unifying Vision*, Cambridge University Press: Cambridge, UK, **2014**.

Carey, FA; Sundberg, RJ. *Advanced Organic Chemistry, Part B: Reactions and Synthesis*, 5th ed., Springer: New York, pp. 1164–1166, **2007**.

Chaisson, EJ. A unifying concept for astrobiology. *International Journal of Astrobiology* **2003**, *2*(2), 91–101. https://doi.org/10.1017/S1473550403001484.

Chaisson, EJ. Energy flows in low-entropy complex systems. *Entropy* **2015**, *17*(12), 8007–8018. https://doi.org/10.3390/e17127857.

Checinska Sielaff, A; Smith, SA. Habitability of Mars: How welcoming are the surface and subsurface to life on the red planet? *Geosciences* **2019**, *9*(9), 361. https://doi.org/10.3390/geosciences9090361.

Chela-Flores, J. From systems chemistry to systems astrobiology: Life in the universe as an emergent phenomenon. *International Journal of Astrobiology* **2013**, *12*(1), 8–16. https://doi.org/10.1017/S1473550412000262.

Chen, IA; de Vries, MS. From underwear to non-equilibrium thermodynamics: Physical chemistry informs the origin of life. *Physical Chemistry Chemical Physics* **2016**, *18*(30), 20005–20006. https://doi.org/10.1039/C6CP90169G.

Chuang, HY; Hofree, M; Ideker, T. A decade of systems biology. *Annual Review of Cell and Developmental Biology* **2010**, *26*,721–744. https://doi.org/10.1146/annurev-cellbio–100109–104122.

Churchill, RP. *Becoming Logical, An Introduction to Logic*, St. Martin's Press: New York, pp. 106–121, **1986**.

Churchill, RP. *Logic, An Introduction*, 2nd ed., St. Martin's Press: New York, **1990**, pp. 122–140.

Chyba, CF; Sagan, C. Comets as a source of prebiotic organic molecules for the Early Earth. In *Comets and the Origin and Evolution of Life*, Thomas, PJ; Chyba, CF; McKay, CP, Eds., 2nd ed., Springer: Heidelberg, Germany, pp. 147–174, **1997**.

Clark, BC. Barriers to natural interchange of biologically active material between Earth and Mars. *Origins of Life and Evolution of Biospheres* **1986**, *16*, 410–411. https://doi.org/10.1007/BF02422102.

Clark, BC. Planetary interchange of bioactive material: Probability factors and implications. *Origins of Life and Evolution of the Biosphere* **2001**, *31*, 185–197. https://doi.org/10.1023/A:1006757011007.

Clark, BC. Martian meteorites do not eliminate the need for back contamination precautions on sample return missions. *Advances in Space Research* **2002**, *30*, 1593–1600. https://doi.org/10.1016/S0273–1177(02)00481–7.

Clark, BC. Searching for extraterrestrial life in our Solar System. In *Handbook of Astrobiology*, Kolb, VM, Ed., CRC Press: Boca Raton, pp. 801–817, **2019a**.

Clark, BC. A generalized and universalized definition of life applicable to extraterrestrial environment. In *Handbook of Astrobiology*, Kolb, VM, Ed., CRC Press: Boca Raton, pp. 65–74, **2019b**.

Clark, BC; Kolb, VM. Macrobiont: Cradle for the origin of life and creation of a biosphere. *Life* **2020**, *10*(11), 278. https://doi.org/10.3390/life10110278.

Clark, BC; Kolb, VM; Steele, A, et al. Origin of life on Mars: Suitability and opportunities. *Life* **2021**, *11*(6)539. https://doi.org/10.3390/life11060539.

Cleaves, HJ. Prebiotic chemistry: Geochemical context and reaction screening. *Life* **2013**, *3*(2), 331–343. https://doi.org/10.3390/life3020331.

Cockell, CS; Bush, T; Bryce, C, et al. Habitability: A review. *Astrobiology* **2016**, *16*(1), 89–117. https://doi.org/10.1089/ast.2015.1295.

Cohen, SM; Curd, P; Reeve, CDC, Eds., *Readings in Ancient Greek Philosophy*, Hackett Publ.: Indianapolis, pp. 42–47, **2000**.

Colín-García, M; Villafañe-Barajas, S; Camprubí, A; Ortega-Gutiérrez, F; Colás, V; Negrón-Mendoza, A. Prebiotic chemistry in hydrothermal vent systems. In *Handbook of Astrobiology*, Kolb, VM, Ed., CRC Press: Boca Raton, FL, pp. 297–330, **2019**.

Complex system. Accessed October 1, 2022. https://en.wikipedia.org/wiki/Complex_system.

Conley, CA. Planetary protection. In *Handbook of Astrobiology*, Kolb, VM, Ed., CRC Press: Boca Raton, FL, pp. 819–834, **2019**.

Constable, DJC; Jimenéz-Gonzáles, C; Matlin, SA. Navigating complexity using systems thinking in chemistry with implications for chemistry education. *Journal of Chemical Education* **2016**, *96*, 2689–2699. https://doi.org/10.1021/acs.jchemed.9b00368

Criado-Reyes, J; Bizzarri, BM; Garcia-Ruiz, JM; Saladino, R; Di Mauro, E. The role of borosilicate glass in Miller-Urey experiment. *Scientific Reports* **2021**, *11*, 21009. https://doi.org/10.1038/s41598-021-00235-4.

Crick, F. *What Mad Pursuit; A Personal View of Scientific Discovery*, Basic Books: New York, **1988**.

Crick, FHC; Orgel, LE. **1973**. Directed Panspermia. *Icarus* **1973**, *19*(3), 391-346. https://doi.org/10.1016/0019–1035(73)90110–3.

Dalai, P; Sahai, N. Protocell emergence and evolution. In *Handbook of Astrobiology*, Kolb, VM, Ed., CRC Press: Boca Raton, FL, pp. 491–520, **2019**.

Deamer, D. Membrane compartments in prebiotic evolution. In *The Molecular Origins of Life: Assembling Pieces of the Puzzle*, Brack, A., Ed., Cambridge University Press: Cambridge, UK, pp. 189–205, **2000**.

Deamer, D. The role of lipid membranes in life's origin. *Life* **2017**, *7*(1), 5. https://doi.org/10.3390/life7010005.

Deamer, DW. *Assembling Life. How Can Life Begin on Earth and Other Habitable Planets?* Oxford University Press: Oxford, UK, **2019**.

Deamer, DW. *Origin of Life: What Everyone Needs to Know*, Oxford University Press: Oxford, UK, **2020**.

De Duve, C. *Blueprint for a Cell: The Nature and Origin of Life*, Neil Patterson Publ.: Burlington, North Carolina, pp. 4–5, **1991**.

Des Marais, DJ. Astrobiology goals: NASA strategy and European roadmaps. In *Handbook of Astrobiology*, Kolb, VM, Ed., CRC Press: Boca Raton, FL, pp. 15–26, **2019**.

Des Marais, DJ; Nuth III, JA; Allamandola, LJ, et al. The NASA astrobiology roadmap. *Astrobiology* **2008**, *8*(4), 715–730. https://doi.org/10.1089/ast.2008.0819.

Eigen, M. *From Strange Simplicity to Complex Familiarity, A Treatise on Matter, Information, Life and Thought*, Oxford University Press: Oxford, UK, **2013**.

Fry, I. Philosophical aspects of the origin-of-life question: Neither by chance nor by design. In *Handbook of Astrobiology*, Kolb, VM, Ed., CRC Press: Boca Raton, FL, pp. 109–124, **2019a**.

Fry, I. The origin of life as an evolutionary process: Representative case studies. In *Handbook of Astrobiology*, Kolb, VM, Ed., CRC Press: Boca Raton, FL, pp. 437–462, **2019b**.

Gayon, G. Defining life: Synthesis and conclusions. *Origins of Life and Evolution of the Biosphere* **2010**, *40*, 231–244. https://doi.org/10.1007/s11084-010-9204–3.

González, R; Butković, A; Escaray, FJ; Martínez-Latorre, J; Melero, Í; Pérez-Parets, E; Gómez-Cadenas, A; Carrasco, P; Elena, SF. Plant virus evolution under strong drought conditions results in a transition from parasitism to mutualism. *Proceedings of the National Academy of Sciences* **2021**, *118*(6). https://doi.org/10.1073/pnas.2020990118.

Gupta, A. Definitions. In *The Stanford Encyclopedia of Philosophy*, Zalta, EN, Ed., Summer 2015 Edition, **2015**. https://plato.stanford.edu/archives/sum2015/entries/definitions/.

Guttenberg, N; Virgo, N; Chandru, K; Scharf, C; Mamajanov, I. **2017**. Bulk measurements of messy chemistries are needed for a theory of the origins of life. *Philosophical Transactions of the Royal Society A* **2017**, *375*, 20160337. http://dx.doi.org/10.1098/rsta.2016.0347.

Hall, JE; Hall, ME. *Guyton and Hall Textbook of Medical Physiology*, Elsevier: Amsterdam, **2021**.

Harold, FM. *The Way of the Cell: Molecules, Organisms and the Order of Life*, Oxford University Press: Oxford, UK, p. 254, **2001**.

Harrison, SA; Lane, N. Life as a guide to prebiotic nucleotide synthesis. *Nature Communications* **2018**, *9*(1), 1–4. https://doi.org/10.1038/s41467-018-07220–y.

Heraclitus. *Heraclitus Fragments*. Translated by Brooks Haxton, Penguin Books: New York, **2001**.

Higgs, PG. When is a reaction network a metabolism? Criteria for simple metabolisms that support growth and division of protocells. *Life* **2021**, *11*(9), 966. https://doi.org/10.3390/life11090966.

Hoffmann, PM. *Life's Rachet: How Molecular Machines Extract Order from Chaos*, Basic Books: New York, p. 87, **2012**.

Horneck, G; Bäcker, H. Can microorganisms withstand the multistep trial of interplanetary transfer? Considerations and experimental approaches. *Origins of Life and Evolution of the Biosphere* **1986**, *16*(3), 414–415. https://doi.org/10.1007/BF02422104.

Horneck, G; Walter, N; Westall, F, et al. AstRoMap European astrobiology roadmap. *Astrobiology* **2016**, *16*, 201–243. https://doi.org/10.1089/ast.2015.1441.

Hud, NV; Fialho, DM. RNA nucleosides built in one prebiotic pot. *Science* **2019**, *366*, 32–33. https://doi.org/10.1126/science.aaz1130.

Ideker, T; Galitski, T; Hood, L. A new approach to decoding life: Systems biology. *Annual Review of Genomics and Human Genetics* **2001**, *2*(1), 343–372. https://doi.org/10.1146/annurev.genom.2.1.343.

Islam, S; Powner, MW. Prebiotic systems chemistry: Complexity overcoming clutter. *Chem* **2017**, *2*(4), 470–501. https://doi.org/10.1016/j.chempr.2017.03.001.

Jeancolas, C; Malaterre, C; Nghe, P. Thresholds in origin of life scenarios. *Iscience* **2020**, *23*(-11), 101756. https://doi.org/10.1016/j.isci.2020.101756.

Kasser, J. *Holistic Thinking: Creating Innovative Solutions to Complex Problems*, The Right Requirement Publishers: Cranfield, UK, pp. 358–367, **2013**.

Kauffman, SA. The adjacent possible. In *Life*, Brockman, J, Ed., Harper Perennial: New York, **2016**.

Kindermann, M; Stahl, I; Reimold, M; Pankau, WM; von Kiedrowski, G. Systems chemistry: Kinetic and computational analysis of a nearly exponential organic replicator. *Angewandte Chemie* **2005**, *117*(41), 6908–6913. https://doi.org/10.1002/ange.200501527.

Kitano, H., Ed. *Foundations of Systems Biology*, The MIT Press: London, England, **2001**.

Kitano, H. Systems biology: A brief overview. *Science* **2002**, *295*(5560), 1662–1664. https://doi.org/10.1126/science.1069492.

Kolb, VM. On the applicability of the principle of the quantity-to-quality transition to chemical evolution that led to life. *International Journal of Astrobiology* **2005**, *4*(3–4), 227–232. https://doi.org/10.1017/S1473550405002818.

Kolb, VM. On the applicability of the Aristotelian principles to the definition of life. *International Journal of Astrobiology* **2007**, *6*(1), 51–57. https://doi.org/10.1017/S1473550407003564.

Kolb, VM. On the applicability of dialetheism and philosophy of identity to the definition of life. *International Journal of Astrobiology* **2010**, *9*, 131–136. https://doi.org/10.1017/S1473550410000017.

Kolb, VM. On the laws for the emergence of life from abiotic matter. In *Instruments, Methods, and Missions for Astrobiology XV*, vol. 8521, p. 852109, **2012**.

Kolb, VM. Development of the algorithm for life for the search for extraterrestrial life. In *Instruments, Methods, and Missions for Astrobiology XVI*, International Society for Optics and Photonics, vol. 8865, p. 88650B, **2013**. https://doi.org/10.1117/12.2021040.

Kolb, VM, Ed., *Astrobiology: An Evolutionary Approach*, CRC Press: Boca Raton, FL, **2015**.

Kolb, VM. Origins of life: Chemical and philosophical approaches. *Evolutionary Biology* **2016**, *43*(4), 506–515. https://doi.org/10.1007/s11692-015-9361-4.

Kolb, VM, Ed., *Handbook of Astrobiology*, CRC Press: Boca Raton, FL, **2019a**.

Kolb, VM. Astrobiology: Definition, scope and a brief overview. In *Handbook of Astrobiology*, Kolb, VM, Ed., CRC Press: Boca Raton, FL, pp. 3–13, **2019b**.

Kolb, VM. Defining life: Multiple perspectives. In *Handbook of Astrobiology*, Kolb, VM, Ed., CRC Press: Boca Raton, FL, pp. 57–64, **2019c**.

Kolb, VM. Oparin's coacervates. In *Handbook of Astrobiology*, Kolb, VM, Ed., CRC Press: Boca Raton, FL, pp. 483–490, **2019d**.

Kolb, VM. Prebiotic reactions in water, "on water", in superheated water, solventless, and in the solid state. In *Handbook of Astrobiology*, Kolb, VM, Ed., CRC Press: Boca Raton, FL, pp. 331–340, **2019e**.

Kolb, VM; Clark III BC. *Astrobiology for a General Reader: A Questions and Answers Approach*, Cambridge Scholars Publishing: New Castle Upon Tune, UK, **2020**.

Kolb, VM; Liesch, PJ. Abiotic, biotic, and in-between. In *Instruments, Methods, and Missions for Astrobiology XI*, International Society for Optics and Photonics, vol. 7097, p. 70970A, **2008**. https://doi.org/10.1117/12.792668.

Kompanichenko, VN. *Thermodynamic Inversion: Origin of Living Systems*, Springer International Publishing, **2017**.

Kompanichenko, V. Thermodynamic jump from prebiotic microsystems to primary living cells. *Sci* **2020**, *2*, 14. https://doi.org/10.3390/sci2010014.

Korpela, EJ. SETI: Its goals and accomplishments. In *Handbook of Astrobiology*, Kolb, VM, Ed., CRC Press: Boca Raton, FL, pp. 727–739, **2019**.

Koskella, B. Resistance gained, resistance lost: An explanation for host–parasite coexistence. *PLoS Biology* **2018**, *16*(9), e3000013. https://doi.org/10.1371/journal.pbio.3000013.

Krishnamurthy, R. Giving rise to life: Transition from prebiotic chemistry to protobiology". *Accounts of Chemical Research* **2017**, *50*, 455–459. https://doi.org/10.1021/acs.accounts.6b00470.

Krishnamurthy, R. Life's biological chemistry: A destiny or destination starting from prebiotic chemistry? *Chemistry—A European Journal* **2018**, *24*(63), 16708–16715. https://doi.org/10.1002/chem.201801847.

Lakdawalla, E. *The Design and Engineering of Curiosity: How the Mars Rover Performs Its Job*, Springer: Berlin, Germany, **2018**.

Lang, C; Lago, J; Pasek, M. Phosphorylation on the early Earth: The role of phosphorus in biochemistry and its bioavailability. In *Handbook of Astrobiology*, Kolb, VM., Ed., CRC Press: Boca Raton, FL, pp. 361–370, **2019**.

Larralde, R; Robertson, MP; Miller, SL. Rates of decomposition of ribose and other sugars: Implications for chemical evolution. *Proceedings of the National Academy of Sciences* **1995**, *92*(18), 8158–8160. https://doi.org/10.1073/pnas.92.18.8158.

Lovelock, J. *The Ages of Gaia: A Biography of our Living Earth*, Norton: New York, **1995**.

Ludlow, RF; Otto, S. Systems chemistry. *Chemical Society Reviews* **2008**, *37*(1), 101–108. https://doi.org/10.1039/B611921M.

Longstaff, A. *Astrobiology: An Introduction*, CRC Press: Boca Raton, FL, **2015**.

Luisi, PL. *The Emergence of Life: From Chemical Origins to Synthetic Biology*, Cambridge University Press: Cambridge, pp. 17–37, **2006**.

Luisi, PL. *The Emergence of Life: From Chemical Origins to Synthetic Biology*, 2nd ed., Cambridge University Press: Cambridge, pp. 119–156. 247–261, **2016**.

Mars Rover website. https://mars.nasa.gov/mars 2020/spacecraft/rover/.

Mason, SF. *Chemical Evolution, Origins of the Elements, Molecules and Living Systems*, Oxford University Press: Oxford, **1991**.

Maturana, HR; Varela, FJ. Autopoiesis and cognition: The realization of the living. In *Boston Studies in the Philosophy of Science*, Vol. 42, Cohen, R.S and Wartofsky, MW., Eds., Springer: Dodrecht, 1980.

Meadows, DH. *Thinking in Systems: A Primer*, Chelsea Green Publishing: White River Junction, Vermont, **2008**.

Melosh, HJ. Impact ejection, spallation and the origin of meteorites. *Icarus* **1984**, *59*, 234–260. https://doi.org/10.1016/0019–1035(84)90026–5.

Melosh, HJ. The rocky road to panspermia. *Nature* **1988**, *332*(6166), 687–688. https://doi.org/10.1038/332687a0.

METI (Messaging Extraterrestrial Intelligence). https://en.wikipedia.org/wiki/METI.

Mileikowsky, C; Cucinotta, FA; Wilson, JW; Gladman, B; Horneck, G; Lindegren, L; Melosh, J; Rickman, H; Valtonen, M; Zheng, JQ. Natural transfer of viable microbes in space: 1. From Mars to Earth and Earth to Mars. *Icarus* **2000**, *145*(2), 391–427. https://doi.org/10.1006/icar.1999.6317.

Miller, SL. A production of amino acids under possible primitive earth conditions. *Science* **1953**, *117*, 528–529.

Miller, SL; Orgel, LE. *The Origins of Life on the Earth*, Prentice Hall: Englewood Cliffs, NJ, **1974**.

National Academies of Sciences, Engineering, and Medicine. *An Astrobiology Strategy for the Search for Life in the Universe*, The National Academies Press: Washington, DC, **2019**.

O'Connor, J; McDermott, I. *The Art of Systems Thinking*, Thorsons: San Francisco, **1997**.

O'Leary, MR. *Anaxagoras and the Origin of Panspermia Hypothesis*, iUniverse, Inc.: New York, **2008**.

Oparin, AI. *The Origin of Life* (published in Russian) in **1924**. English translation by A. Sygne published in *Origins of life: The Central Concepts*, Deamer, DW; Fleischaker, GP, Eds., Jones and Bartlett: Boston, MA, pp. 31–71, **1994**.

Oparin, AI. *Origin of Life*, **1938**. Translated by S. Morgulis. Republication of the original publication by Macmillan Company by Dover Publications Inc., 2nd ed., Dover: New York, **1965**.

Oparin, AI. *Genesis and Evolutionary Development of Life*, published in Russian in **1966**, English translation by Maass, E., Academic Press: New York, **1968**.

Oren, A. Extremophiles and their natural niches on Earth. In *Handbook of Astrobiology*, Kolb, VM, Ed., CRC Press: Boca Raton, pp. 635–660, **2019**.

Orgill, MK; York, S; MacKellar, J. Introduction to systems thinking for the chemistry education community. *Journal of Chemical Education* **2019**, *96*(12), 2720–2729. https://doi.org/10.1021/acs.jchemed.9b00169.

Padgett, JF; Powell, WW. *The Emergence of Organizations and Markets*, Princeton University Press: Princeton, NJ, **2012**.

Pályi, G; Zucci, C; Caglioti, L, Eds. *Fundamentals of Life*, Elsevier: Amsterdam, pp. 15–55, **2002**.

Parkin, KLG. **2018**. The Breakthrough Starshot system model. *Acta Astronautica* **2018**, *152*, 370–384. https://doi.org/10.1016/j.actaastro.2018.08.035.

Pasek, MA; Kee, TP; Bryant, DE; Pavlov, AA; Lunine, JI. Production of potentially prebiotic condensed phosphates by phosphorus redox chemistry. *Angewandte Chemie Int Ed* **2008**, *47*(41), 7918–7920. https://doi.org/10.1002/anie.200802145.

Pasek, MA. Thermodynamics of prebiotic phosphorylation. *Chemical Reviews* **2020**, *120*(11), 4690–4706. http://dx.doi.org/10.1021/acs.chemrev.9b00492.

Patel, BH; Percivalle, C; Ritson, DJ; Duffy, CD; Sutherland, JD. Common origins of RNA, protein and lipid precursors in a cyanosulfidic protometabolism. *Nature Chemistry* **2015**, *7*, 301–307.

Peretó, J. Controversies on the origin of life. *International Microbiology* **2005**, *8*, 23–31.

Peretó, J. Out of fuzzy chemistry: From prebiotic chemistry to metabolic networks. *Chemical Society Reviews* **2012**, *41*, 5394–5403. https://doi.org/10.1039/C2CS35054H.

Peretó, J. Prebiotic chemistry that led to life. In *Handbook of Astrobiology*, Kolb, VM, Ed., CRC Press: Boca Raton, pp. 219–233, **2019**.

Perry, RS; Kolb, VM. On the applicability of Darwinian principles to chemical evolution that led to life. *International Journal of Astrobiology* **2004**, *3*(1), 45–53. https://doi.org/10.1017/S1473550404001892.

Popa, R. *Between Necessity and Probability: Searching for the Definition and Origin of Life*, Springer-Verlag: Heidelberg, pp. 197–205, **2004** (definitions of life are given in chronological order starting from 1855).

Popa, R. Necessity, futility and the possibility of defining life are all embedded in its origin as a punctuated-gradualism. *Origins of Life and Evolution of Biospheres* **2010**, *40*, 183–190. https://doi.org/10.1007/s11084-010-9198-x.

Popa, R. Elusive definition of life: A survey of main ideas. In *Astrobiology: An Evolutionary Approach*, Kolb, VM, Ed., CRC Press: Boca Raton, pp. 325–348, **2015**.

Powner, MW; Sutherland, JD. Prebiotic chemistry: A new modus operandi. *Philosophical Transactions of the Royal Society B* **2011**, *366*(1580), 2870–2877. https://doi.org/10.1098/rstb.2011.0134.

Powner, MW; Sutherland, JD; Szostak, JW. The origins of nucleotides. *Synlett* **2011**, *14*, 1956–1964. https://doi.org/10.1055/s-0030-1261177.

Priest, G. *Beyond the Limits of Thought*, Oxford University Press: Oxford, UK, **2002**.

Priest, G. What is so bad about contradictions? In *The Law of Non-Contradiction: New Philosophical Essays*, Priest, G; Beall, JC; Armour-Garb, B, Eds., Oxford University Press: Oxford, UK, pp. 23–38, **2006**.

Priest, G; Berto, F; Weber, Z. *Dialetheism, The Stanford Encyclopedia of Philosophy*, Zalta, EN, Ed., Fall 2018 ed., **2018**. https://plato.stanford.edu/archives/fall2018/entries/dialetheism/.

Prigogine, I. *The End of Certainty*, The Free Press: New York, **1997**.

Prigogine, I; Nicolis, G; Babloyantz, A. Thermodynamics of evolution. The functional order maintained within living systems seems to defy the Second Law; nonequilibrium thermodynamics describes how such systems come to terms with entropy. *Physics Today* November **1972a**, 23–28.

Prigogine, I; Nicolis, G; Babloyantz, A. Thermodynamics of evolution. The idea of nonequilibrium order and of the search for stability extend Darwin's concept back to the prebiotic stage by redefining the 'fittest'. *Physics Today* December **1972b**, 38–44.

Prigogine, I; Stengers, I. *Order out of Chaos*, Bantam Books: New York, **1984**.

Pross, A. The driving force for life's emergence: Kinetic and thermodynamic considerations. *Journal of Theoretical Biology* **2003**, *220*(3), 393–406.

Pross, A. How does biology emerge from chemistry? *Origins of Life and Evolution of Biospheres* **2012**, *42*(5), 433–444.

Pross, A. *What is Life? How Chemistry Becomes Biology*, 2nd ed., Oxford University Press: Oxford, UK, pp. 163–164, **2016**.

Ragsdale, SW; Pierce, E. Acetogenesis and the Wood–Ljungdahl pathway of CO_2 fixation. *Biochimica et Biophysica Acta (BBA)-Proteins and Proteomics* **2008**, *1784*(12), 1873–1898. https://doi.org/10.1016/j.bbapap.2008.08.012.

Ranjan, S; Sasselov, DD. Influence of the UV environment on the synthesis of prebiotic molecules. *Astrobiology* **2016**, *16*(1), 68–88. https://doi.org/10.1089/ast.2015.1359.

Ranjan, S; Todd, Z; Sutherland, JD; Sasselov, DD. Sulfidic anion concentrations on early Earth for surficial origins-of-life chemistry. *Astrobiology* **2018**, *18*(8), 1023–1040.

Repko, AF. *Interdisciplinary Research: Process and Theory*, 2nd ed., Sage: Thousand Oaks, CA, pp. 20–22, 73, 94–96, **2012**.

Ricard, J. Systems biology and the origins of life? Part I. Are biochemical networks possible ancestors of living systems? Reproduction, identity and sensitivity to signals of biochemical networks. *Comptes Rendus Biologies* **2010a**, *333*(11–12), 761–768. https://doi.org/10.1016/j.crvi.2010.10.004.

Ricard, J. Systems biology and the origins of life? Part II. Are biochemical networks possible ancestors of living systems? Networks of catalysed chemical reactions: Non-equilibrium, self-organization and evolution. *Comptes Rendus Biologies* **2010b**, *333*(11–12), 769–778. https://doi.org/10.1016/j.crvi.2010.10.004

Ricard, J. Systems biology and the origins of life. *Current Synthetic and Systems Biology* **2013**, 2, 3. https://doi.org/10.4172/2332-0737.1000e110.

Ricard, J. Emergence of information and the origins of life: A tentative physical model. *Current Synthetic and Systems Biology* **2014a**, 2, 1. https://doi.org/10.4172/2332-0737.1000e102.

Ricard, J. *Biological Systems: Complexity and Artificial Life*, Bentham Science Publishers, pp. 15–34, 185–195, May 6 **2014b**. https://benthambooks.com/.

Ridley, M. *Genome; The Autobiography of a Species in 23 Chapters*, Harper Collins Publishers: New York, NY, p. 15, **1999**.

Ritson, DJ; Battilocchio, C; Ley, SV; Sutherland, JD. Mimicking the surface and prebiotic chemistry of early Earth using flow chemistry. *Nature Communications* **2018**, *9*(1), 1–10. https://doi.org/10.1038/s41467-018-04147-2.

Ritson, DJ. A cyanosulfidic origin of the Krebs cycle. *Science Advances* **2021**, *7*(33), eabh3981. https://doi.org/10.1126/sciadv.abh3981.

Rivera-Valentín, EG; Filiberto, J; Lynch, KL, et al. Introduction—First billion years: Habitability. *Astrobiology* **2021**, *21*(8), 893–905. https://doi.org/10.1089/ast.2020.2314.

RNA hydrolysis: Wikipedia RNA. https://en.wikipedia.org/wiki/RNA_hydrolysis.

Robbins, SJ; Hynek, BM. A new global database of Mars impact craters ≥ 1 km: Database creation, properties, and parameters. *Journal of Geophysical Research: Planets* **2012**, *117*, E5. https://doi.org/10.1029/2011JE003966.

Ross, DS; Deamer, D. Dry/wet cycling and the thermodynamics and kinetics of prebiotic polymer synthesis. *Life* **2016**, *6*, 28. https://doi.org/10.3390/life6030028.

Ruiz-Mirazo, K; Briones, C; de la Escosura, A. Prebiotic systems chemistry: New perspectives for the origins of life. *Chemical Reviews* **2014**, *114*, 285–366. https://doi.org/10.1021/cr2004844.

Sandford, SA; Brownlee, DE; Zolensky, ME. The Stardust sample return mission. In *Sample Return Missions, Last Frontiers of Solar System Exploration*, Longobardo, A, Ed., Elsevier: Amsterdam, pp. 79–104, **2021**.

Sasselov, DD, Grotzinger, JP, Sutherland, JD. The origin of life as a planetary phenomenon. *Science Advances* **2020**, *6*(6), eaax3419. https://doi.org/10.1126/sciadv.aax3419.

Schmitt-Kopplin, P; Gabelica, Z; Gougeon, RD; Fekete, A; Kanawati, B; Harir, M; Gebefuegi, I; Eckel, G; Hertkorn, N. High molecular diversity of extraterrestrial organic matter in Murchison meteorite revealed 40 years after its fall. *Proceedings of the National Academy of Sciences of the USA* **2010**, *107*, 2763–2768. https://doi.org/10.1073/pnas.0912157107.

Schmitt-Kopplin, P; Harir, M; Kanawati, B; Gougeon, R; Moritz, F; Hertkorn, N; Clary, S; Gebefugi, I, Gabelica, Z. Analysis of extraterrestrial organic matter in Murchison meteorite: A progress report. In *Astrobiology: An Evolutionary Approach*, Kolb, VM, Ed., CRC Press: Boca Raton, FL, pp. 63–82, **2015**.

Schulze-Makuch, D; Irwin, LN. *Life in the Universe: Expectations and Constraints*, Springer Verlag: Heidelberg, **2004**.

Smith, HH; Hyde, AS; Simkus, DN, et al. The grayness of the origin of life. *Life* **2021**, *11*(6), 498. https://doi.org/10.3390/life11060498.

Solé, R; Elena, SF. *Viruses as Complex Adaptive Systems*, Princeton University Press: Princeton, **2019**.

Spirkin, A; Yakhot, O. *The Basic Principles of Dialectic and Historical Materialism*, Progress, Moscow, pp. 49–51, **1971**.

Spitzer, J. The complexity of life's origins: A physicochemical view. In *Handbook of Astrobiology*, Kolb, VM, Ed., CRC Press: Boca Raton, FL, pp. 463–482, **2019**.

Strazewski, P. The beginning of systems chemistry. *Life* **2019a**, *9*(1), 11. https://doi.org/10.3390/life9010011.

Strazewski, P. The essence of systems chemistry. *Life* **2019b**, *9*(3), 60. https://doi.org/10.3390/life9030060.

Strazewski, P. Prebiotic chemical pathways to RNA and the importance of its compartmentation. In *Handbook of Astrobiology*, Kolb, VM, Ed., CRC Press: Boca Raton, FL, pp. 235–264, **2019c**.

Strazewski, P. Coevolution of RNA and peptides. In *Handbook of Astrobiology*, Kolb, VM, Ed., CRC Press: Boca Raton, FL, pp. 409–420, **2019d**.

Sutherland, JD. The origin of life—Out of the blue. *Angewandte Chemie Int Ed* **2016**, *55*(1), 104–121. https://doi.org/10.1002/anie.201506585.

Sutherland, JD. Studies on the origin of life-The end of the beginning. *Nature Reviews Chemistry* **2017**, *1*, 0012. https://doi.org/10.1038/s41570-016-0012.

Systems Chemistry: Wikipedia. https://en.wikipedia.org/wiki/Systems_chemistry.

Systems Biology: Wikipedia. https://en.cikipedia.org/wiki/Systems_biology.

Systems Thinking and Practice, The Open University: Milton Keynes, UK, The Kindle Edition, **2016**.

Szostak, JW. Systems chemistry on early Earth. *Nature* **2009**, *459*, 171–172. https://doi.org/10.1038/459171a.

Szostak, JW. The narrow road to the deep past: In search of the chemistry of the origin of life. *Angewandte Chemie Intl Ed* **2017**, *56*, 1107–11043. https://doi.org/10.1002/anie.201704048.

Tavassoly, I; Goldfarb, J; Iyengar, R. Systems biology primer: The basic methods and approaches. *Essays in Biochemistry* **2018**, *62*(4), 487–500. https://doi.org/10.1042/EBC20180003.

Tirard, S. The debate over panspermia: The case of the French botanists and plant physiologists at the beginning of the twentieth century. In *The History and Philosophy of Astrobiology: Perspectives on Extraterrestrial Life and the Human Mind*, Dunér, D; Parthemore, J; Persson, E; Holmberg, G., Eds., Cambridge Scholars Publishing: Newcastle upon Tyne, pp. 213–221, **2013**.

Todd, ZR; Öberg, KI. Cometary delivery of hydrogen cyanide to the early Earth. *Astrobiology* **2020**, *20*(9), 1109–1120. https://doi.org/10.1089/ast.2019.2187.

Treiman, AH; Gleason, JD; Bogard, DD. The SNC meteorites are from Mars. *Planetary and Space Science* **2000**, *48*(12–14), 1213–1230. https://doi.org/10.1016/S0032–0633(00)00105–7.

Van Helmont, JB. All life is chemistry, 1648. In *Genome; The Autobiography of a Species in 23 Chapters*, Ridley, M, Ed., Harper Collins Publishers: New York, NY, p. 15, **1999**.

Villarreal, LP. Are viruses alive? *Scientific American* **2004**, *291*(6), 100–105. Accessed July 10, 2021. http://www.jstor.org/stable/26060805.

Villarreal, LP; Witzany, G. Social networking of quasi-species consortia drive virolution via persistence. *AIMS Microbiology* **2021**, *7*(2), 138–162. https://doi.org/10.3934/microbiol.2021010.

Virus (Virus – Wikipedia). https://en.wikipedia.org/wiki/Virus.

Vitas, M; Dobovišek, A. Towards a general definition of life. *Origins of Life and Evolution of Biospheres* **2019**, *49*, 77–88. https://doi.org/10.1007/s11084-019-09578-5.

Voet, D; Voet, JG. *Biochemistry*, 4th ed., Wiley: Hoboken, NJ, **2011**.

Voit, EO. *Systems Biology, A Very Short Introduction*, Oxford University Press: Oxford, UK, **2020**.

Von Bertalanffy, L. *General Systems Theory*, Revised ed., George Braziller: New York, **1993**.

Von Hegner, I. *An ab initio definition of life pertaining to Astrobiology*, **2019**. https://hal.archives-ouvertes.fr/hal–02272413v2.

Von Kiedrowski, G; Otto, S; Herdewijn, P. Welcome home, systems chemists! *Journal of Systems Chemistry* **2010**, *1*(1). https://doi.org/10.1186/1759–2208–1–1.

Walker, SI. Transition from abiotic to biotic: Is there an algorithm for it? In *Astrobiology: An Evolutionary Approach*, Kolb, VM, Ed., CRC Press: Boca Raton, FL, pp. 371–379, **2015**.

Wing, MR; Bada, JL. Geochromatography on the parent body of the carbonaceous chondrite. *Geochimica et Cosmochimica Acta* **1991**, *55*(10), 2937–2942. https://doi.org/10.1016/0016–7037(91)90458–H.

Witzany, G. Language and communication as universal requirements for life. In *Astrobiology: An Evolutionary Approach*, Kolb, VM, Ed., CRC Press: Boca Raton, FL, pp. 349–369, **2015**.

Witzany, G. What is Life? *Frontiers in Astronomy and Space Sciences* **2020**, *7*, 7. https://doi.org/10.3389/fspas.2020.00007.

Wołos, A; Roszak, R; Żądło-Dobrowolska, A; Beker, W; Mikulak-Klucznik, B; Spólnik, G; Dygas, M; Szymkuć, S; Grzybowski, BA. Synthetic connectivity, emergence, and self-regeneration in the network of prebiotic chemistry. *Science* **2020**, *369*(6511). https://doi.org/10.1126/science.aaw1955.

Wörmer, L; Hoshino, T; Bowles, MW, et al. Microbial dormancy in the marine subsurface: Global endospore abundance and response to burial. *Science Advances* **2019**, *5*(2), eaav1024.

Wu, LF; Sutherland, JD. Provisioning the origin and early evolution of life. *Emerging Topics in Life Sciences* **2019**, *3*(5), 459–468. https://doi.org/10.1042/ETLS20190011.

Xue, Q; Miller-Jensen, K. Systems biology of virus-host signaling network interactions. *BMB Reports* **2012**, *45*(4), 213–220. https://doi.org/10.5483/BMBRep.2012.45.4.213.

Yadav, M; Kumar, R; Krishnamurthy, R. Chemistry of abiotic nucleotide synthesis. *Chemical Reviews* **2020**, *120*(11), 4766–4805. https://doi.org/10.1021/acs.chemrev.9b00546.

York, S; Orgill, M. ChEMIST table: A tool for designing or modifying instruction for a systems thinking approach in chemistry education. *Journal of Chemical Education* **2020**, *97*(8), 2114–2129. https://doi.org/10.1021/acs.jchemed.0c00382.

Yu, H; Su, Z; Chaput, JD. Darwinian evolution of an alternative genetic system provides support for TNA as an RNA progenitor. *Nature Chemistry* **2012**, *4*(3), 183–187. https://doi.org/10.1038/nchem.1241.

Zhang, SJ; Duzdevich, D; Ding, D; Szostak, JW. Freeze-thaw cycles enable a prebiotically plausible and continuous pathway from nucleotide activation to nonenzymatic RNA copying. *Proceedings of the National Academy of Sciences* **2022**, *119*(17), e2116429119. https://doi.org/10.1073/pnas.2116429119.

Zimmer, C. *Life's Edge: The Search for What It Means to Be Alive*, Dutton, an imprint of Penguin Random House, LLC, UK, pp. 270–271, **2021**.

Zubay, G. *Origins of Life on the Earth and the Cosmos*, 2nd ed., Harcourt Academic Press, San Diego, **2000**.

Index

Note: **Bold** page numbers refer to tables and *italic* page numbers refer to figures.

Printed in Great Britain
by Amazon

34007497R00097